Luminos is the Open Access monograph publishing program from UC Press. Luminos provides a framework for preserving and reinvigorating monograph publishing for the future and increases the reach and visibility of important scholarly work. Titles published in the UC Press Luminos model are published with the same high standards for selection, peer review, production, and marketing as those in our traditional program. www.luminosoa.org

Sonic Socialism

Sonic Socialism

Crisis and Care in Pandemic Hanoi

———

Christina Schwenkel

UNIVERSITY OF CALIFORNIA PRESS

University of California Press
Oakland, California

Suggested citation: Schwenkel, C. *Sonic Socialism: Crisis and Care
in Pandemic Hanoi*. Oakland: University of California Press, 2025.
DOI: https://doi.org/10.1525/luminos.249

Cataloging-in-Publication Data is on file at the Library of Congress.

ISBN 978-0-520-42327-5 (cloth)
ISBN 978-0-520-41619-2 (pbk.)
ISBN 978-0-520-41620-8 (ebook)

GPSR Authorized Representative: Easy Access System Europe,
Mustamäe tee 50, 10621 Tallinn, Estonia, gpsr.requests@easproject.com

34   33   32   31   30   29   28   27   26   25
10   9   8   7   6   5   4   3   2   1

CONTENTS

List of Figures and Media Files — vi

Acknowledgments — ix

Preface: The Pandemic as Critical Sound Event — xiii

Pandemic Figure 1. Superspreaders: Patient 17 (Bệnh Nhân 17) — 1

Introduction: The *Other* COVID Exception — 7

Pandemic Figure 2. Essential Workers: The Security Guard (Người Bảo Vệ) — 24

Act 1. Quarantine: State Care — 29

Pandemic Figure 3. Domestic Workers: The Housekeeper (Người Giúp Việc) — 57

Act 2. Lockdown: Self-Care — 62

Pandemic Figure 4. Nonessential Workers: The Grab Driver (Anh Grab) — 94

Act 3. Social Distancing: Mutual Care — 98

Pandemic Figure 5. Migrant Workers: The Philippine Singer (Ca Sĩ Philippines) — 133

Coda: Epi(demi)c Endings — 137

Notes — 149

References — 177

Index — 195

FIGURES

1. Still from the video "Ghen Cô Vy," or "COVID Envy"   *xv*
2. Lockdown of Trúc Bạch Street   *2*
3. Temperature screening with acoustic feedback   *27*
4. Health e-declaration for entry into Vietnam   *31*
5. Immigration hall becomes COVID testing site   *37*
6. Empty shelves in California supermarket   *40*
7. Stocked shelves in Hanoi supermarket   *44*
8. Food distribution during collective quarantine   *47*
9. Police sound truck doubles as public health vehicle   *48*
10. Home medical visit from health station   *52*
11. Morning sweep to care for the commons   *59*
12. "Fighting a Pandemic Is Like Fighting the Enemy"   *65*
13. Loudspeaker attached to utility pole   *71*
14. Loudspeakers mounted on cultural center   *72*
15. Audio relics of war at nostalgia café   *75*
16. *The Red Days III* silk painting   *76*
17. Pandemic solidarity stamps   *77*
18. Supplemental portable speaker system   *83*
19. Alleyway acoustics and use of the commons   *89*
20. Human-nonhuman animal encounters   *101*
21. Trúc Bạch Lake walking path   *101*
22. Bypassing barricades to access park amenities   *102*

23. Still from "Việt Nam Ơi! Đánh Bay Covid" ("Hey, Vietnam! Let's Fight Covid")   *105*
24. Youth pioneer cutout promotes health mandates   *108*
25. Soundmapping of lakeside sonic ecology   *113*
26. Itinerant vendors return to Trúc Bạch Lake   *122*
27. Gendered soundscapes of outdoor fitness   *123*
28. Hip hop COVID-19 street art   *140*
29. COVID-19 public health poster   *147*

VIDEOS

1. Decontamination after residential medical screening   *53*
2. Lockdown evening stroll   *93*
3. Massage chain in public park   *111*
4. Caged birds singing by the lake   *116*
5. Early morning cardio workout   *124*
6. Ghost town soundwalk along Tạ Hiện Beer Street   *128*

AUDIO CLIPS

1. Collective morning sweeping ritual   *59*
2. Live reading of health screening notice on public loudspeaker   *80*
3. State messaging competes with ambient sounds   *83*
4. Construction during lockdown   *91*
5. Nonhuman sonic ecology: interspecies dialogue   *115*
6. Mobile police loudspeaker unit   *119*
7. *Tào phớ* vendor's shifting vocalizations   *121*
8. Kinetic-powered amplification of *bánh đa kê* vendor   *122*
9. Mid-Autumn Festival soundscape   *132*

ACKNOWLEDGMENTS

In a time when care—and its absence—is frequently invoked, acknowledgments offer a small but meaningful gesture of relational caretaking in recognition of those who cared enough to listen, comment, collaborate, or critique. My deepest gratitude extends to the people who live, work, and socialize around Trúc Bạch Lake, whose urban care and carefulness both during and beyond the pandemic made this neighborhood truly vibrant and sonically evolving throughout the years my spouse and I called it home from 2017 to 2024.

This book emerged more by serendipity than by design. When my neighbor—Vietnam's Patient 17—and later I myself became entangled in the thick bureaucracy of pandemic caretaking, these encounters unexpectedly opened space to contemplate the unevenly shared vulnerabilities at the center of this autoethnography. I did not initially imagine this material taking shape as a monograph, however. In fact, during mandatory isolation, I was deep in the copy edits and page proofs of my previous work, *Building Socialism: The Afterlife of East German Architecture in Urban Vietnam*, when these experiences began to coalesce.

The genesis of this sensory autoethnography is indebted to Zeynep Gürsel, whose intellectual generosity and friendship have consistently encouraged me to bring my ideas into the public sphere. Through virtual conversations, as I recounted the sights and sounds of my daily life, Zeynep urged me to write an op-ed piece that reframed the narrative of Vietnam—not as a "synonym for US occupation" but as a space of hope, a place "outside of COVID time." That initial idea gradually expanded into a vision for a larger book project, which found its champion in Kate Marshall at UC Press. Kate's enthusiasm and thoughtful editorial guidance throughout the process were instrumental in transforming my listening practices into this autoethnography.

Telling different stories about Vietnam has always presented challenges, however, and this book is no exception. How do we honor both the spirit of care and its troubling failures while respecting people's agency in striving toward a better future? The generous feedback I received from Tine Gammeltoft and two anonymous readers as part of the peer review process helped me navigate these questions, pushing me to develop more nuanced perspectives throughout this work. I remain deeply grateful to them and the entire team at UC Press who worked with me to bring these ideas to fruition in open-access form through the Luminos program.

This book has greatly benefitted from ongoing dialogue with colleagues across Vietnam. My collaboration with the Hà Nội Ad Hoc architectural collective, particularly during our workshop on multisensory cities in fall 2023, expanded my thinking in new directions. My thanks go to founder and creative force Mai Hưng Trung, along with Đức Lê, Nguyễn Mạnh Trí, Nguyễn Vũ Hải, Phạm Thị Thu Trang, and Ylan Vo. Thu Trang and Hải were especially intrepid companions during our urban explorations. I am also indebted to Phạm Phương Chi, Nguyễn Thị Phương Châm, Cao Thảo Hương, Nguyễn Thế Sơn, Ray Mallon, Linda Mazur, and Suzanne Lecht for their thoughtful conversations and embodied participation.

My Fulbright year in Hồ Chí Minh City (2023–2024) was enriched by the support of colleagues at the University of Social Sciences and Humanities who collaborated on my sensory cities project—especially Trương Thị Thu Hằng, chair of anthropology, long-time friend Trần Thị Thảo, and Cô Phúc and her students. My ideas were also shaped by animated conversations with fellow Fulbrighters Alisa Freedman, Jonathan Nashel, and Rebecca Brittenham, who generously shared her syllabus on urban walking as a cultural lens for literary analysis. Exchanges with Phan Lâm Nhật Nam of K59 Atelier, Gary Paige, Afra Rebuscini and Yuri Frassi of Officine Gặp, and Hiếu Trương of Tản Mạn Kiến Trúc offered new approaches for thinking about sonic life and architectural form. Graduate students in the field with me, especially Shani Tra, Darius Sadighi, Chari Hamratanaphon, Quynh Truong, Tiên Dung Ha, Ichi Ha, Zhiyi Wang, and Tara Westmor, also provided important insights. Thank you to Tiên Dung and her mother for the sensory tour of Cư xá Thanh Đa. Chari's work on the images and Quynh's work on the sound files proved invaluable.

Although this book was researched and written in Vietnam, colleagues in Europe and the United States brought new dimensions to its analysis. I am thankful for ongoing collaborations and discussions with Sarah Grant, Anne Marie Leshkowich, Deborah Wong, Zeynep Gürsel, Liisa Malkki, Sophea Seng, Tine Gammeltoft, Amy Dao, Annuska Derks, Diane Fox, Erik Harms, Julie Chu, Susan Ossman, Sara Swenson, Tâm Ngô, and Paul Ryer. Both Deborah Wong and Jennifer Hsieh have inspired my thinking about sound and its relation to care. In Berlin, my research and writing process was enhanced through conversations with Jennifer Ruth Hosek, Sung Tieu, Claudia Borchers, Alice von Bieberstein, Gertrud Hüwelmeier, Agnieszka Joniak-Lüthi, and Andrea Muehlebach. I thank Otto Stuparitz for his thoughtful comments as discussant for the EuroSEAS 2024 panel "Sounding Power and Dissent in Southeast Asia" and for leading our group on a tour of the University of Amsterdam's Decolonizing Southeast Asian Sound Archives (DeCoSEAS), which proved especially productive for thinking about imperial soundscapes.

Sections of this book were presented at various stages of development to audiences at George Washington University, UC San Diego, the University of Chicago, the University of North Carolina at Chapel Hill, UC Santa Cruz, and Dartmouth College. I extend my sincere thanks to the organizers and attendees at these events for their generative questions and thoughtful responses, including Sarah Wagner, Julie Chu, Kamala Russell, Ryan Cecil Jobson, Christian Lentz, Becky Butler, Chad Bryant, Curt Gambetta, Sujin Eom, Yiman Wang, Boreth Ly, and Hunter Bivens. Early in the writing process, members of the SoCal Southeast Asian Studies collective at UC San Diego—including Claire Edington, Jody Blanco, Mohammad Khamsya Bin Khidzer, Phung Su, Cindy Nguyen, Drew Trinidad, and Diu-Huong Nguyen—offered valuable feedback that helped shape the direction of the project in its formative stages.

My family continues to be a source of care and creative inspiration. During a time when families experienced devastating losses, I was fortunate that my loved ones remained in relatively good health. Across those anxious years, the support of my spouse was unwavering. This work was born of difficult circumstances, and I remain thankful that we were able to be together during the pandemic, when many transnational families faced long separations. From assisting with sound collection to walking alongside me—both literally and figuratively—my spouse offered steady care and companionship, grounding this project in the very practices of caretaking it seeks to highlight.

Preface

The Pandemic as Critical Sound Event

On February 23, 2020, Vietnam's Ministry of Health released a YouTube video titled "Ghen Cô Vy," or "COVID Envy." The playful music animation, which features humans encountering mischievous microbes, unexpectedly became an international sensation. Dubbed the "Handwashing Song" in English, it quickly racked up tens of millions of views worldwide.[1] What began as a public service announcement (PSA) about a rogue coronavirus exhibiting human-like traits of agency and intentionality ironically went viral itself. The video transcended its original context, capturing the attention of global audiences, including in the United States. Its international prominence grew even further after being featured on the television program *Last Week Tonight*, where host John Oliver's enthusiastic endorsement ignited another wave of online buzz. His comedic performance and effusive praise for the "incredible" video, as he termed it, resonated strongly, especially in Vietnam. Local media there proudly reported on the positive international publicity, further amplifying the video's impact.

"Ghen Cô Vy" sought to render the virus identifiable through sonic cues like coughing and wheezing, while instilling proper hygiene practices. As an example of medical knowing through embodied sensing, the video sparked lively discussions in my Anthropology of Southeast Asia class at the University of California at Riverside. Released in early 2020, when Vietnam had confirmed only a handful of infections, the video exemplified the country's preemptive response and its readiness to take swift action well before the pandemic's full impact was felt. Its strategic rollout foreshadowed the government's innovative use of sound media, including audiovisual formats, as cornerstones of national public health planning aimed at disseminating information, galvanizing collective action, and curbing the transmission of disease. This multimodal approach to emergency preparedness yielded significant results during the critical early months of COVID-19, with Vietnam successfully staving off any pandemic-related fatalities until July 31, a feat matched by only a handful of nations worldwide.[2]

"Ghen Cô Vy" became an earworm that was difficult to shake. The song's catchy refrain and cute set of characters with their danceable moves sparked a nationwide trend in Vietnam: the *vũ điệu rửa tay*, or "handwashing dance." The music's infectious melody and hygiene-centered choreography would quickly find new life as a TikTok global dance challenge. With the hashtag #GhenCoVyChallenge, dancers worldwide improvised the hand-lathering and scrubbing sequence, syncing their moves with the lyrics while adding creative twists. This viral phenomenon altered the original purpose of the PSA, as dancers infused the hygiene routine with their own distinctive styles and cultural expressions, while expanding its reach and influence.

The core message of "Ghen Cô Vy" highlights the importance of knowledge and an informed public as essential foundations for building and maintaining collective resilience against the pandemic. At the center of the story are two human protagonists—a male and female of unspecified age and relationship—who diligently execute a series of science-based hygiene rituals to defend themselves from a menacing coronavirus. This nonhuman antagonist, a spiky-green entity donning a yellow crown, bares fierce eyes and sharp fangs, leaving little doubt about its malicious intent to inflict harm.

The impulse to anthropomorphize, or ascribe human characteristics to nonhuman actants, has long been an artistic practice in graphic medical narratives (Saji, Venkatesan, and Callender 2021, 151; Gui 2024). In "Ghen Cô Vy," this technique manifests in the coronavirus's inflated sense of purposeful agency. The pathogen is personified through its given name (Corona) and designated place of origin or *quê* (Wuhan), framing it as a foreign adversary. This narrative taps into a familiar historical trope in Vietnam of mobilizing the population against a shared, external threat.[3] The cheerful melody serves as a striking counterpoint to the virus's cartoonishly villainous qualities. Its exaggerated, human-like traits amplify both hostile emotion and murderous motivation, underscoring the gravity of the public health emergency through contrasting visual and auditory elements. The playful yet urgent tone balances this intensity while urging viewers to adopt heightened— and even hyper—sensory vigilance (figure 1).

As an energetic musical tutorial, "Ghen Cô Vy" exemplifies the Vietnamese government's innovative use of interactive media and sonic storytelling to distill complex information about infectious disease into simple, memorable sound bites. With clever lyrics and animation to "raise social awareness" (*nâng cao ý thức của xã hội*), the video features an ensemble of frontline figures who debunk misinformation with scientific facts. The narrative champions a unified pandemic response, emphasizing

FIGURE 1. Standing guard to the whimsical tune of "COVID Envy": a human barricade faces down a swarm of aggressive coronaviruses. Still from "Ghen Cô Vy," 2020.

that triumph over the virus depends not on luck or superstition, but on the consistent adoption of embodied, evidence-based hygiene practices. Teaming up with artists, public health officials transformed mundane acts of self-care, like handwashing, into enjoyable communal activities through an instructional PSA masking as a Vietnamese pop song (V-pop). With lyrics such as "Let's wash our hands together, rub rub rub rub all over" ("Cùng rửa tay xoa xoa xoa xoa đều"), the soundtrack fosters a sense of belonging and shared purpose, linking individual compliance with collective well-being through "decisive self-action" (*quyết tâm tự giác*). This musicking of collectivity (Small 1998) orchestrates an uplifting message of compassionate care and moral duty in uniting to safeguard the country.

"Ghen Cô Vy" concludes on a hopeful note, declaring, "Our Vietnam is determined to prevail over the disease" ("Việt Nam ta quyết thắng bệnh dịch"), where the use of *ta* (our) reinforces a spirit of communal ownership and civic responsibility in preventing the spread of COVID-19. The final scene juxtaposes raised fists and beating hearts against the national flag—affectively charged imagery that harmonizes science, caregiving, and nationalism through song.[4] This audiovisual synergy transforms pandemic precautions from state impositions into resonant expressions of Vietnamese solidarity, elevating hygiene habits into markers of ethical citizenship.

Vietnam boasts a rich cultural history of utilizing folk songs and didactic poetry for public health education. During the revolutionary period of cultural reform, these sonic traditions were reworked into civilizing practices aimed at cultivating modern hygienic citizens. The use of auditory techniques to teach about disease prevention parallels what Laurence Coderre identifies in the context of the Chinese Cultural Revolution: a drive to "produce the nation as a sonic space of mass publicity and socialist modernity" (2021, 21).

The Vietnamese state revived and amplified its acoustic traditions during the COVID-19 pandemic, a "critical event" that thrust society into "new and unprecedented terrains" (Das 1995, 5). Rooted in anticolonial efforts to "sonically territorialize the nation" (Coderre 2021, 21), these sounding practices evolved to address contemporary health care challenges through a media infrastructure encompassing both digital and nondigital platforms. What I term "sonic socialism" in this book connects sound and listening practices ("sonic") to ways of organizing and governing society ("socialism"). This concept captures how sound has remained a vital tool in public health governance, serving as a mechanism for both exercising state power and instilling collective discipline to safeguard human life during national crises.

During the pandemic, sounds came to supplant images to assume what Julian Henriques (2011) calls the "sonic dominance" of the urban environment. Because containment measures that restricted movement rendered street-level visual messaging less effective, sonic strategies emerged as persuasive means for communicating risk and managing biosecurity. To govern the population remotely during periods of isolation, the state devised a repertoire of sounding practices that blended innovation with historical precedent.[5] This autoethnography examines these audio media technologies and their associated practices of embodied listening, analyzing how they operated both independently of and in tandem with other sensory modalities, as exemplified by the music video. Through this lens, the pandemic manifested as a critical sound event that projected authority and care-driven messaging into the intimate spaces of homes. At the same time, as the TikTok dance challenge revealed, listeners actively repurposed state-produced sounds and directives, underscoring the inherently reciprocal and relational nature of sonic governance.

. . .

"One Year Ago," read major newspaper headlines on March 11, 2021, marking the anniversary of the World Health Organization's (WHO) declaration that COVID-19, caused by the rapidly spreading SARS-CoV-2 virus, had become a global pandemic.[6] The WHO's urgent announcement, which highlighted the severity of the disease, prompted much of the world to shut down, including the United States. The one-year anniversary of the declaration evoked a flood of emotional responses in the US media, as people recounted the pivotal moments, sounds, artefacts, and affects that defined their experiences of immense social disruption and isolation brought on by the unprecedented public health emergency. Vivid recollections of panic buying and empty shelves in anticipation of a catastrophe as global supply chains collapsed formed archives of pandemic memories that centered around seemingly ordinary sounds, like synchronized clapping, and commonplace commodities, such as toilet paper, which took on particular symbolic significance. Trivial documents, including hastily scrawled shopping lists and somber workplace closure emails that foreshadowed the dramatic shifts in life to come, transformed into meaningful relics of a world upended. These immersive chronicles of the present carried an affective urgency that powerfully conveyed the shared sentiments of dread, shock, anguish, and loss, making tangible the notion that "being in history" is inherently a "densely corporeal" experience (Berlant 2011, 64). Rather than an ephemeral past that "flits by" or "flashes up" in a transitory moment of danger (Benjamin 1969, 255), this was a deeply felt, richly sensorial past, collectively lived and acutely sensed over the protracted unfolding of pandemic time—a duration that far exceeded initial expectations. During that first year in the United States, the haunting wail of sirens and stark visuals of outbreaks—graphs and diagrams tracing cyclical "waves" and sudden "spikes" alongside the quest for "peaks" and "flattened curves"—demanded a readjustment of expectations and a recalibration of sensory orientations as society faced the grim reality that the virus would not be swiftly contained.

I, too, found myself immersed in this relational "affectsphere" (Berlant 2011, 64) that enveloped the world, grappling with the emergent threat of infectious disease. This threat catalyzed acts of solidarity across scales—from local mutual aid networks to international arrangements of "mask diplomacy" that temporarily upended global hierarchies. However, disparities and racial scapegoating soon surfaced. With deepening concern, I bore witness to the bare supermarket shelves—a foreboding sign of the scarcity of capitalism that loomed—and the racism exhibited toward Asian youth blamed for buying *too many* masks, tensions that would only escalate. In my own desperate search for personal protective equipment (PPE), I eventually secured one of the last remaining boxes of alcohol swabs, tucked away behind a pharmacy counter to control access. On March 10, while teaching my final class of the quarter on Southeast Asian ethnography, I, too, received "the email"—that fateful communication suspending in-person instruction. The university's closure, the email read with misplaced optimism, would only last through April 3.

Throughout the course, the students and I had observed the gradual spread of COVID-19 across the region of Southeast Asia. We watched and hummed along to Vietnam's upbeat "handwashing song," appreciating its melody and choreography cleverly designed to educate the public about virus prevention measures. The thought that our lives would be so abruptly and profoundly disrupted by the quarter's end was inconceivable.[7] Instruction had commenced just days before the WHO announced on January 9 the detection of a novel coronavirus in Wuhan, China. The first reported case in the United States emerged on January 21—two days prior to Vietnam's first case on January 23—during the third week of classes. As the quarter progressed into its ninth week by early March, the US nationwide tally stood at a seemingly manageable thirty cases. Just ten days later, when the directive to transition to remote teaching was issued, the reported number of infections had escalated thirtyfold, nearing one thousand cases across the country.[8]

And yet, early indications of an imminent breakdown of public health infrastructure—and a broader failure of governing institutions in the United States—were apparent from the outset. Despite long-standing warnings about the inevitability of infectious disease emergencies (Lakoff 2017, 7), no comprehensive national strategy for biosecurity existed to mitigate the spread of infections. In his March 11, 2020, address declaring the crisis a bona fide pandemic, WHO Director-General Tedros Adhanom Ghebreyesus urged nations to respond proactively by adopting a "whole-of-government, whole-of-society approach" in order to "save lives and minimize impact."[9] However, the United States failed to mount a coordinated response. Vietnam, on the other hand, implemented swift and comprehensive public health measures. For instance, whereas a lack of testing capacity impeded prompt detection and containment of the virus in the United States, Vietnam took early precautionary actions for testing, tracing, and isolating as part of its "Zero-COVID" strategy.[10] This approach proved strikingly effective: Vietnam, with a population of ninety-seven million, was the only country of its size to report no COVID-19 deaths six months into the pandemic (Haider et al. 2020, 5–6). Moreover, the US public showed an alarming lack of knowledge and concern regarding the virus, in contrast to Vietnam's well-informed and vigilant population.

The United States' unpreparedness at multiple levels of outbreak response is made starkly evident in my correspondence with a graduate student evacuated from her research site in Italy as the virus swept through Europe. An Asian American Fulbright fellow living under strict lockdown in Lombardy (Italy's epicenter), she received escalating instructions from the Fulbright Commission, first to leave the region, then a week later to return to California before the month-long suspension of all travel to and from Europe began on March 13. Upon arriving at Los Angeles International Airport (LAX) and disclosing that she had been living in what the media dubbed the "deadliest coronavirus hotspot on the planet," the immigration officer casually waved her through. In a March 14 email sent from self-imposed hotel quarantine (undertaken to protect her family, including elderly parents), she reported that none of her evacuated Fulbright peers across the United States had been tested upon return. "I [asked] if I could be tested for the purposes of preventing community transmission," she wrote, "but they told me 'welcome back' and that [was] it. . . . In Italy, everyone was being screened."

My student's pandemic experience shifted from witnessing a militarized state of emergency in Italy—with troops patrolling the streets—to encountering inaction and even skepticism in the United States. Her subsequent search for a testing facility in Riverside County proved exasperating. When she contacted the student health center, she was referred to urgent care, where staff refused to test her without symptoms, despite her insistence that she could be asymptomatic. She was sternly warned that any attempt to feign illness to access screening would be detected through her temperature reading, underscoring how Asian individuals were often subject to heightened scrutiny and suspicion.

My own efforts to help also ended in frustration. Though the campus had recently launched a twenty-four-hour COVID-19 hotline for faculty and staff, my messages over two days went unreturned. When I reached out to Riverside County health officials, they directed me to a website with a toll-free number. My student called, hoping to schedule a test, but the facility had already reached its full capacity. When she explained the urgency of the situation and the financial burden of self-quarantine, they provided a list of shelters as a potential resource, seemingly oblivious to the risk of community spread among vulnerable populations in emergency housing. After four days, she finally secured a test appointment. She then spent another week in hotel quarantine, financial strain mounting, anxiously awaiting the results.

As the country descended into chaos, becoming the world's deadliest COVID-19 hotspot—with a death toll exceeding one hundred thousand by late May 2020, just three months after the first recorded US fatality—I began making plans to leave. While I recognize the privilege of being able to make choices about my mobility as a white cisgender woman with stable employment, which made me less vulnerable to the virus than more marginalized racial and economic groups, I could sense disaster looming, and I knew that the window for departure was rapidly closing. Amid widespread global travel bans, my university responded to the escalating crisis by announcing travel restrictions effective at midnight on March 17.

I had intended to travel to Vietnam over spring break to start a new research project and reunite with my spouse in Hanoi. Flight cancellations stoked my fear of indefinite separation, a reality countless transnational families would soon face. Given Vietnam's proven track record of decisive public health interventions during zoonotic outbreaks (Porter 2019), I was confident it would take robust, preemptive action against COVID-19. The music video "Ghen Cô Vy," though a lighthearted animation, offered a compelling glimpse into Vietnam's disaster preparedness, particularly its capacity to rapidly coordinate and mobilize resources.

Michel Foucault proposed the concept "eventalization" to challenge the notion that events unfold through a self-evident, universal chain of causality (1991, 76–77). This idea proves valuable for reexamining the March 11, 2020, declaration of a global pandemic. Rather than treating this declaration as a "historical constant . . . which imposes itself uniformly," eventalization prompts a reflection on what truly counts as a defining juncture worthy of being marked and remembered (1991, 76). The designation of COVID-19 as a global pandemic marked a watershed moment in the United States—solemnly commemorated one year later—but held divergent meaning elsewhere in the world. Vietnam, for instance, only declared its "nationwide epidemic" on April 1, a full three weeks after the WHO pronouncement, illustrating how pandemic time unfolded differently in each country, shaped by culturally specific norms and national priorities. As the anthropologist Marshall Sahlins (1985) astutely observed, a phenomenal occurrence only becomes an "event" through interpretation, thereby bearing a "distinctive cultural signature" (1985, xviii–xiv). Despite media portrayals suggesting a universal structure of feeling—embodied by the masked global citizen as a symbol of shared humanity (Barthes 1972, 100–2)—COVID-19 exposed vast disparities in how people around the world perceived and responded to the pandemic.

Of course, each nation experienced its own unique pandemic timeline, with outbreak temporalities emerging as pivotal events shaped by local contexts. For Vietnam, the first critical turning point came on January 23, when it identified its first COVID-19 cases—a man and his son from Wuhan—launching the country's "fight" (*cuộc chiến*) against the novel coronavirus and inspiring the creation of the PSA jingle "Ghen Cô Vy." One year later, the Vietnamese press extensively covered this "eventalized" anniversary, highlighting the initial infections and the country's early achievements in containing the virus. By January 2021, Vietnam had recorded only 1,548 cases and just thirty-five deaths, reflecting one of the lowest pandemic mortality rates globally at that moment.

No less consequential was Vietnam's critical outbreak event on March 6 originating in Hanoi with Patient 17 (Bệnh Nhân 17), the country's first jet-setter super-spreader (see pandemic figure 1). This first cluster of COVID-19 infections in the densely populated capital city triggered a decisive shift in the government's emergency response, with strict measures like business closures aimed at curbing potential community transmission within crowded urban neighborhoods. The outbreak hit close to home: Patient 17 was my neighbor in Trúc Bạch Ward, Ba Đình District, Hanoi. Authorities moved quickly to lock down and secure the surrounding area, confining an entire street block of sixty-six families to their homes for two weeks. This pivotal development heightened my sense of urgency to depart the United States before any quarantine expansion to our adjacent lakeside residence, which fortunately never occurred, or before Vietnam closed its borders, which it did on March 25. As I headed to the airport amid reports of surging cases and overwhelmed medical facilities in the United States—laying bare the failures of capitalism in times of crisis (Adams 2020)—I had a gnawing sense that I would not return to California by the end of spring break.

A surreal calm pervaded the international terminal at LAX, devoid of the usual teeming crowds and bustling atmosphere. The passengers boarding my flight to Taipei, many of whom appeared to be international students, seemed driven by the same urgency I myself felt to leave the country and reunite with family before the window of opportunity slammed shut. A palpable tension hung in the air. Passenger jets were widely considered key conduits for the airborne transmission of the virus, a concern fueled by the cautionary tale of Vietnam's Patient 17, which made international headlines. Adding to the tension, several evening flights to

Asia had already been cancelled, casting a haze of uncertainty over the status of my flight and subsequent connection to Hanoi.

Pandemic developments at that time were rapidly evolving worldwide, changing by the hour. Anxious about potential contagion or flight cancellation, I boarded the Boeing 777 with trepidation and quickly found my seat. A wave of relief washed over me as the door sealed shut. With less than a third of its seats occupied, the plane pulled away from the gate, taxied across the tarmac, and took to the skies just before midnight. Despite cancellations throughout Taipei's international airport, my connecting flight to Vietnam remained on schedule. I made my way through the eerily empty terminal to continue the onward journey, joined by only a handful of fellow passengers—all Vietnamese nationals—on the nearly vacant plane bound for Hanoi.

Pandemic Figure 1

Superspreaders

Patient 17 (Bệnh Nhân 17)

By all accounts, it was a typical Friday evening in Trúc Bạch Ward in Ba Đình District, Hanoi, on March 6, 2020. The cafés and eateries lining the narrow streets and winding their way around Trúc Bạch Lake, where John McCain's plane had been shot down during the war, were bustling with the usual activity. The mood was upbeat and hopeful. There had been no new COVID-19 cases reported in over three weeks, and all sixteen previously infected patients from the past two months had recovered. It was early morning in California, and my spouse and I were on a FaceTime call as he headed out the door to see one of our favorite bands—a group of Vietnamese musicians who had studied in Czechoslovakia in the 1980s and performed Pink Floyd covers they had learned overseas. We were planning my arrival for the following week, growing apprehensive about the coronavirus spreading beyond Asia and across Europe after its rapid escalation in Lombardy, Italy. A few hours later, my phone pinged with a series of instant messages, then rang. "There's been an outbreak," my spouse said anxiously as I viewed the texted images of police and medical personnel in protective suits behind a blockade, signaling the shift from pandemic preparedness to an active state of emergency (figure 2). "In our neighborhood."

The Trúc Bạch outbreak (*sự bùng phát*) marked a critical rupture in the rhythm of everyday urban life, as its "spectacular eventfulness" (Caduff 2014, 41) plunged the city into an eerie stillness. As the first recorded cases in Hanoi, the crisis catalyzed a pivotal change in the state's pandemic policy and approach to care governance, altering the course of action from mere anticipation of infection to aggressive interventions to contain the virus at any cost. Comprising a sequence of linked infection events, the outbreak also marked the appearance of

FIGURE 2. Spectacular eventfulness: lockdown of several dozen households on Trúc Bạch Street, March 2020. Photo by the author.

Vietnam's earliest "superspreader" (*bệnh nhân siêu lây nhiễm*), known as Patient 17 (Bệnh Nhân 17).

Quantifying and tracking COVID-19 infections through numerical identifiers during the pandemic's first year reflected a politics of enumeration that structured both disease surveillance and public awareness of the virus's embeddedness in the social fabric. These closely watched numbers and their real-time updates, as part of a broader infrastructure of calculation (Murphy 2017), granted authorities legitimacy through the reassuring effect that countability offered to an anxious public. The comparatively slow rise in the daily reported infection counts not only demonstrated the efficacy of containment measures but also showcased the state's commitment to transparency and accountability. Beyond mere statistics, these numeric designations conferred a peculiar notoriety, usurping individual identities, as exemplified by the case of Patient 17, whose illness and contagion became the subject of intense national scrutiny and international epidemiological study.

The March 6 outbreak produced both fear and fascination with the privileged jet-setter at the center of the story: twenty-six-year-old NHN, as she was initially identified in the press, the daughter of a steel magnate, who carried the virus back from Europe on her eleven-hour flight from London to Hanoi.[1] This marked the first known occurrence of long-haul, atmospheric spread of the airborne disease.[2] Meticulous contact tracing found that NHN had infected at least fifteen other passengers, most of whom were seated near her in business class. These infected passengers then traveled to all corners of Vietnam and beyond, including to her family (aunt) and household staff.

Hanoi's sensational outbreak narrative contained all the elements of a conventional epidemiological account of disease emergence and spread. Such outbreak narratives, as Priscilla Wald (2008, 2) argues, function as paradigmatic stories with far-reaching moral, political, and social consequences, capable of either promoting or mitigating the stigmatization of individuals or groups and their conduct. Moreover, these narratives shape how people, including scientists, understand and imagine the threat of infection. Public references to SARS-CoV-2 as the "Wuhan virus," for example, drew on historical fears of a "yellow peril" through persistent racist tropes of Asians as vectors of disease, which fueled anti-Asian violence and xenophobia in US society, Marie Myung-Ok Lee (2020) observed as early as February in her commentary in *Salon*.[3] COVID-19 Orientalism served as a sober reminder of the precarious boundary between epidemiological profiles and racial or ethnic profiling (Kleinman and Lee 2005, 182).

Historically, disease—and disease carriers—have been as gendered and classed as they have been racialized. The outbreak narrative that unfolded around socialite and influencer NHN had much in common with the archetypal story of the healthy *female* disease carrier—that of Mary Mallon, or Typhoid Mary as she became known. Like Typhoid Mary, NHN's biomedical transformation into Patient 17 raised critical questions about what Wald (2008, 73) describes as the relationship between "social responsibility and bacteriological individualism"— namely, the extent to which individuals were accountable not only for their own health and those with whom they had contact, but also for the well-being of the social body, now at grave risk. Both Typhoid Mary and Patient 17 highlighted the dangers of permeable borders in an interconnected world. Yet while outbreak narratives typically frame contagion as invasion (by migrant or enemy Others), in this instance, the threat to the nation emerged from within.[4]

Patient 17 became a cautionary tale of a fallen woman whose illness was portrayed as a consequence of reckless, if not illicit, behavior that endangered the lives of innocent others, marking outbreaks as both socio-ethical and biomedical events. This moral framing of the virus, itself mobile and airborne, reinforced a binary perception of purity versus pollution, with the latter often ascribed to unruly female bodies. In a society deeply suspicious of conspicuous wealth, public scrutiny has long focused disapprovingly on women's freedom of movement and lack of supervision. Details circulated about NHN and her sister's lavish lifestyle, including photographs of their front-row attendance at Milan Fashion Week before she boarded the plane to fly home. (Her sister remained in Europe and, soon after, was hospitalized with COVID-19.) Upon arrival in Vietnam, NHN switched passports—she also held British citizenship, casting further doubt on her national loyalty—presumably to hide her recent visit to a viral hotbed in Italy. Allegations surfaced that she had intentionally misled authorities by providing a false statement about her medical status—concealing that she was symptomatic— and her travel history to circumvent the mandatory quarantine imposed on those returning from high-risk areas. Critics drew attention to the conspicuous warning displayed at the top of the health declaration form that air passengers were required to submit, which explicitly stated, "Declaring false information is a violation of Vietnamese law that could result in criminal prosecution" (Khai báo thông tin sai là vi phạm pháp luật Việt Nam, có thể bị xử lý hình sự). Consequently, Patient 17 was perceived by the public as a willful and knowing biosecurity threat.

If disease is a metaphor for social disorder, as Susan Sontag (1978, 58) famously stated, then the superspreader outbreak narrative, refracted through class and gender politics, became an opportunity to express concern about the future of the Vietnamese nation-state. Patient 17 embodied the dangers of capitalist society— its moral decay and rampant materialism that clashed with the collectivist values of socialism, including its logics of care that were ostensibly shared. In this context, Patient 17 emerged as a scapegoat for both social and epidemiological contamination. In placing the female self before society and hedonism before the population's well-being, her viral contagion threatened to erode not only social foundations but also the very basis of national belonging, one premised on the healthy (re: uncorrupted) body that was free from disease.

Patient 17's indulgence in a moment of crisis that called for restraint drew the ire of a panicking public. Particulars about her case swiftly spread online; within an hour, she had been identified.[5] Outraged netizens lambasted the superspreader, targeting her social media accounts for online harassment, which were swiftly shut down. Damaging rumors, accusations, and calls for her imprisonment followed, prompting the deployment of police outside her newly constructed residence—the most imposing structure on the block, visible from the rear of my apartment—as a precautionary measure against potential threats. NHN's public shaming caught the attention of international media and provided fodder for anti-communist critics. The *New Yorker*'s sympathetic portrayal of NHN as a victim of the Vietnamese government's "regular use [of] newspaper leaks to persuade or frighten its citizens" (Max 2020) sparked a backlash in the Vietnamese media for its perceived misinformation. This included the article's assertion that NHN had "accidentally" spread the virus, thus absolving her of any culpability. Vietnamese media proponents also criticized the article for privileging liberal notions of privacy over an ethos of social responsibility, an ethos, they argued, that had enabled Vietnam, a resource-constrained country, to effectively avert a public health catastrophe.[6]

The stigmatization of Vietnam's first superspreader was not only virtual but also spatial. After barriers had been erected at both ends of the street (at Ngũ Xã and Châu Long), Patient 17's property was sealed off with a guard posted outside, marking its social isolation. The grounds remained shrouded in darkness during the neighborhood's fourteen-day quarantine, and from my vantage point, the building appeared unoccupied, devoid of any signs of life. (To this day, it stands vacant, an eerie reminder of those turbulent times.)

The state's containment strategy employed targeted lockdowns and medical surveillance, including twice-daily health monitoring. Under this stringent policy, F0s—those confirmed to be infected—were promptly hospitalized, while F1s—individuals who had been in close contact with F0s—were placed in centralized quarantine facilities. F2s and F3s, on the other hand, were allowed to isolate themselves within the confines of their homes.

Given the dense nature of urban living in Hanoi, outbreaks necessitated the implementation of what might be called "cohort quarantines" within affected areas, encompassing entire city blocks, alleyways, and apartment buildings, from aging socialist-era housing units to sleek, new high-rise condominiums. These zones of containment were secured with a perimeter and staffed with on-site medical practitioners charged with administering care to those within. This embrace of "disaster collectivism" (Grossman 2021), emphasizing communal-based responses to crises rather than individualized solutions, challenged the perception of outbreak events as isolated household emergencies. Instead, it fostered a sense of relational solidarity among impacted communities by governing through principles of mutual care and cooperation, uniting citizens in the face of adversity. Yet beneath the surface of collective action lurked the possibility that those in power had appropriated notions of care and caregiving to secure influence while advancing their personal or political agendas (Hobart and Kneese 2020, 8).

The WHO cautioned that stigma surrounding the coronavirus could weaken social cohesion.[7] Yet the Trúc Bạch outbreak, marking an inflection point in Vietnam's response to COVID-19, seemed to strengthen sentiments of national togetherness as residents and local government joined forces to contain the virus. Neighbors and family members formed support networks, delivering care packages to those in quarantine. The state stepped in to ensure the well-being of a vulnerable population by assuming responsibility for the provision of food, medicine, and preventative health education. Observing daily rituals of resource distribution firsthand (see figure 8), I witnessed how these public enactments embodied core principles of care as both politically and affectively charged practice (Puig de la Bellacasa 2011, 90).

This point resonated again when I heard Đào Vân Anh, the deputy party secretary of Trúc Bạch Ward, remark emotionally, "I felt the essence of humanity" (Tôi thấy được tình người) while describing the coordinated response to the outbreak during her appearance on the Vietnam Television (VTV) program, *Morning Coffee*, after the two-week lockdown was lifted on March 20, with no further infections and no reported deaths. As a celebratory gesture, the secretary, together with the chairman of the Ward People's Committee, and the program hosts, performed the "handwashing dance" (*vũ điệu rửa tay*) to the popular tune of "Ghen Cô Vy," the "COVID Envy" song (see preface). The celebration took an unexpected turn when the composer, Khắc Hưng, made a surprise appearance and joined in the performance, ending with elbow bumps and shoe taps. With a laugh, the secretary reflected on the ward's successful, holistic approach to managing care and preserving life, which included playful sonic strategies: "Of all the meaningful work carried out by leaders and the community, the most memorable was practicing proper handwashing techniques with quarantined residents by dancing to the melody of 'Ghen Cô Vy.'"[8]

Introduction

The Other *COVID Exception*

Pandemics disrupt. As emergent crises, they alter the course of urban futurity and how people perceive, navigate, and make sense of risk and uncertainty in the city. The sensory dimensions of crises framed as disasters demand greater attention, particularly as experienced through sonic rupture. Pandemics not only generate profound social, temporal, and material disruptions to everyday life. They also effect a radical sensory disorientation, one that compels new ways of being in, caring for, and attuning to our urban worlds through sounds.

The concept of the "sonic" provides a framework for examining how these sensory ruptures manifest and shape acoustic responses to crises. By "sonic," I refer to a full auditory spectrum that includes not only sound itself but also the diverse modalities of listening, perceiving, engaging with, and producing sound—practices that are inseparable from other embodied sensory experiences. This framework allows for viewing sound-based practices not merely as technological interventions in crisis management but as a lens for understanding how disaster responses reflect deep cultural values—core values that the Vietnamese state, the subject of this book, considers essential to the survival of socialism.

As the COVID-19 pandemic began its global spread in early 2020, Vietnam faced a pivotal moment. The March 6 outbreak in Hanoi's Trúc Bạch Ward, located in the government district of Ba Đình, became a tipping point in the country's response, prompting an emergency quarantine of an entire neighborhood. This decisive action foreshadowed the nationwide lockdown that followed three weeks later. By the end of April, Vietnam had achieved the unthinkable: containment of the virus with fewer than three hundred infections and no recorded deaths. For the next ninety-nine days, no cases of community transmission were reported, even as infection numbers spiked in the Global North. With closed borders, the country appeared to be a sanctuary from the viral disease ravaging much of the world. This swift containment sparked displays of pride and pandemic patriotism across the population. Yet Vietnam's success received little recognition in media outside of the region. This disconnect led me to ask: What lessons might be drawn from Vietnam's coordinated crisis response? And, furthermore, how might those efforts offer insight into the relationship between power, culture, and the collective will that mobilized national action?

The Trúc Bạch outbreak, the first in Hanoi, activated the state's sweeping pandemic preparedness plan across all scales of urban governance. This set into motion a cascade of coordinated interventions that transformed the city and its soundscapes.[1] That pivotal night, the bustle of urban life came to an abrupt halt. By early the next morning, the streets were empty. Pre-dawn walkers, aerobic enthusiasts, and teams of cyclists had vanished; the usually crowded *phở* stalls stood padlocked and silent. A surreal stillness enveloped the lakeside district, broken only by the authoritative female voice of a *loa phường*, or public loudspeaker, urging residents to stay inside and monitor their health. Fear permeated the air.

Patient 17, identified as the superspreader (see pandemic figure 1), had introduced a state of bio*in*security. This fragile, new reality was wrought by unseen—and unforeseen—vulnerabilities to contagion, border threats, and the risk of economic collapse (Chen and Sharp 2014). When I arrived on the scene to reunite with family ten days later, I found a dramatically altered cityscape. The municipal government had swiftly mandated the closure of all nonessential businesses, and my neighbors had retreated to the safety of their homes. The once-vibrant city lay shrouded in an eerie silence. If certain sounds come to define a city's sonic identity, how might a city redefine itself in the absence of those sounding practices? What role does sound play in governance strategies for risk reduction and care provision in times of crisis? And how might modes and contexts of listening change in response?

Sonic Socialism examines how the Vietnamese state managed a public health crisis through acoustic interventions in everyday life in Hanoi, the nation's capital. Based on socially distant "soundwork" and "footwork" during the first year of the pandemic, as an alternative to conventional fieldwork, I highlight the pivotal yet overlooked role that sound reproduction technologies played in Vietnam's efforts to achieve some of the world's lowest infection and death rates, despite being a high-risk country bordering China. Across lived experiences of state containment policies—spanning quarantine, lockdown, and spatial distancing—I analyze how disaster governance operated sonically as a mechanism of state power, deployed to instruct, manage, and safeguard the population, while framing these actions as "care" to ensure national biosecurity.

In an era dominated by vision, what can the sounds associated with declared emergencies tell us about crisis governance and lived experiences of differential vulnerability? How might attunement to sonic atmospheres reveal new dimensions of states of exception and their implications for collective life? Through embodied practices of sensory autoethnography, this meditation on isolation traces both the state's use of sound to regulate and prevent COVID-19, and the forms of sonic dissent that emerged in response. I develop the concept of "sonic governance of care" to identify the exercise of authority not *over* sound (as in policing noise) but *through* sound to maintain public health and protection from disease.[2] In decentering *urban ocularcentrism*—the dominance of the visual in studying the city—the book calls for greater attention to the sensory disruptions and reorientations that have transformed everyday urban encounters with risk and uncertainty. Shifting from visual to auditory registers, a distinct narrative of the pandemic emerges, one that foregrounds sonic ways of knowing, illuminating the rhythms and dissonances of inter/disconnected life. This approach demonstrates the transformative power of sound to reimagine ways of living, being, sensing, and caring in a city while navigating the complexities of a global health emergency.

COVID-19: WHO *DID* GET IT RIGHT?

On February 24, 2023, the *New Yorker Radio Hour* released a podcast titled "The Pandemic at Three: Who Got It Right," marking the start of COVID-19's fourth year (Remnick 2023). The guest speaker, Dhruv Khullar, a writer for the *New Yorker*, offered his assessment while noting that government responses had evolved as the virus mutated: "The countries that got things right, the ones that, you know, spring to mind, are many East Asian countries." I perked up, wondering if Vietnam would be mentioned. Khullar continued: "So Japan, South Korea, Singapore, Taiwan, um . . . there's also countries like New Zealand and Australia that did very well." Disappointment set in as I realized Vietnam would not be included, even as he elaborated on what those countries—all high-income—had done right. Most importantly, he concluded, successful countries had prioritized early detection and containment of the virus through coordinated actions. These included mass screening, rigorous contact tracing, and the transparent dissemination of real-time information, measures that demonstrated the ability to quickly control outbreaks while cultivating strong social cohesion and trust in institutions. As I listened, I mentally checked off each point for Vietnam.[3] Yet there was no mention of the country, despite its cumulative number of infections over the three-year period being considerably lower than that of several others discussed. Moreover, by the end of the pandemic's first year—the focus of this book—Vietnam had recorded the fewest cases per million among these very nations.[4]

Scholars have argued that Vietnam stands as a global exemplar in infectious disease containment. It was the first country to swiftly eradicate SARS in 2003 (Fahlman 2019), and it halted the spread of avian flu a year later (Porter 2019), bolstering the state's legitimacy both at home and abroad.[5] These interspecies crises afforded Vietnam the opportunity to strengthen its public health infrastructure and biosecurity governance to enhance disaster resilience.[6] Given this history, why was Vietnam's coordinated, multiscalar response to another zoonotic pathogen—SARS-CoV-2—repeatedly overlooked, downplayed, or dismissed as the mere product of authoritarianism? This dismissal denied Vietnamese agency and public expressions of solidarity, even as the country implemented prevention policies that were as rigorous as—if not less restrictive than—those of Asian democracies praised for handling the pandemic effectively.

Socialist societies like Vietnam have long been viewed through a lens of deficiency and lack (Dunn and Verdery 2011, 253). These nations are frequently judged as falling short of the global standards, best practices, and accountability metrics set by Western capitalist economies, a perspective deeply embedded in Cold War histories of anti-communism. This logic of "deficit thinking" projects these standards onto North-South and East-West hierarchies, framing postcolonial countries in particular as needing improvement or even intervention to ensure international biosecurity through the policing of racialized bodies. Narratives of deficiency have deep roots in medical discourses of "colonial pathologies" that portrayed colonies as perilous places of sickness and disease (Anderson 2006; Dewachi 2017). Across (post)socialist and (post)colonial contexts, notions of disorder and "mismanagement," informed by the racial logics of empire, have routinely served as proxies for assessing levels of civilization and moral development, marking those nations that fail to conform to hegemonic imperial norms (Liboiron 2021, 74–75).

The burden of double Otherness haunts Vietnam as an Asian socialist country with a sustained history of anticolonial radicalism.[7] A Cold War Orientalism that once justified US military intervention in Vietnam now informs its strategies of trade integration and regional securitization.[8] Moral allegations of mismanagement—for example, of mismanaged waste, such as plastics that find their way into the ocean or "dumped" shrimp and catfish raised in polluted (mismanaged!) waters that "flood" the US market—have become matters of international security. Rather than fight communism, Americans now fight fish, the latest "yellow peril."[9] Embroiled in racial politics, fish have been weaponized as threats to both the US economy and its public health. In this "new war" against Vietnamese sentient nonhumans, the consumption of "contaminated" seafood is deemed a danger to American bodies, a claim that deflects US culpability for creating the ecocidal conditions in Vietnam that continue to harm diverse forms of life.

COVID-19 upended the global management hierarchy by challenging the notion that societies outside Western modernity are inherently diseased, disorderly, or "in deficit" and thus "need to learn to manage better, or be managed" (Liboiron 2021, 75). Observers noted that many countries in the Global South avoided catastrophic outbreaks during the first year of the pandemic, while (neo)colonial powers became hubs of sickness and death. In other words, poorer countries in many instances managed the crisis more adeptly than wealthier nations with advanced health care systems.[10] This pattern was also evident in late socialist countries, including Cuba, with robust response capabilities, supported by an infrastructure of emergency preparedness and socialized public health (Brotherton 2012).

For all its technological achievements, the Global North, Arjun Appadurai observed, was not at the vanguard of biosecurity management and prevention but rather occupied "the strange position of welcoming assistance from . . . unconventional actors on the world stage" that possessed far fewer resources (2020, 222). In other words, the power dynamics that have historically underpinned conditions of hardship and acts of generosity were unsettled as wealthier countries came to rely on "gifts" from those they viewed as inferior to mitigate their scarcity. This included Vietnam, which donated masks and other medical equipment, including ventilators, to Europe and the United States—as well as regional destinations—to boost its soft power. However, Appadurai does not mention Vietnam on his list of "COVID exceptions," which does include China and Cuba. As the *other* (and Othered) COVID exception, Vietnam's actions upended the entrenched South-North hierarchy between mismanaged beneficiary and well-governed benefactor. This inversion transformed the conceit of US exceptionalism into white American anxiety, exposing its "fantasies of mastery" and revealing—through its spectacular *mis*management—that the world's superpower was unable to protect life from premature death (Ahuja 2016, vii).[11]

The specter of the "Vietnam syndrome"—deployed as a "somber metric" by which all other conflicts and their political outcomes or human losses are measured (S. Nguyen 2020)—haunted this excess mortality and its racial disparities. On April 28, 2020, US media outlets reported that COVID-19 fatalities in the United States had surpassed the number of Americans killed in the war in Vietnam.[12] This false equivalency overlooked both the enormous loss of *Vietnamese* life during the conflict (estimated at three million people) and the protection of life during the pandemic through the quick and coordinated actions taken by the Vietnamese government, which had zero recorded deaths at the time.[13] While nations with mounting death tolls were consumed with ritualized care for the dead—to the extent that it was even possible—Vietnam's biopolitical response centered on maintaining and regulating *life*.

Vietnam's more coherent management of public health and biosecurity troubled the presumptions of nations like the United States, whose stake in global hegemony made such competence difficult to acknowledge. That a supposedly "mismanaged" country, allegedly plagued with economic and ecological deficiencies (like unruly shrimp!) could outperform dominant powers proved challenging to reconcile. This cognitive dissonance manifested in another pandemic iteration of the "Vietnam Syndrome" in the media, which racialized Vietnam's response. A Western tabloid, engaging in explicit COVID racial Othering, characterized the country's public health actions as inherently *inhumane*. The article suggested that the Vietnamese population overcame the virus not through care but through cruelty—specifically, by torturing innocent felines as a folk remedy to prevent viral infection.[14] This portrayal of Vietnam descending into bestiality through improper human-nonhuman relations signified yet another instance of mismanagement, this time the failure to adequately manage human-animal biopolitics (Chen 2012, 134). Similar to Africa (another "bestial" region in imperial imaginaries), Vietnam faced persistent suspicion from Western reporters over allegations of underreporting coronavirus casualties. The minimal case numbers and absence of fatalities, while much of the world suffered terribly, were impossible for many to accept as verifiable truth—so much so that some journalists even contacted funeral homes in Hanoi to disprove official figures (Dao 2024, 146). "How do you know there wasn't a coverup?" skeptics outside Vietnam would ask me, their deep distrust of state-controlled media leading them to dismiss the statistics as "fake news."

An impulse for cooperation and cohesion rather than a condition of coercion also proved to be a difficult construct for outside observers steeped in assumptions about sovereign power and subjugated bodies as the basis for political life in Vietnam. This is not to negate the expansive reach of Vietnamese state and party power through disciplinary interventions aimed at securing the nation from biothreats. But Vietnam was not unique in this regard; as Foucault (1978) reminds us, biopower—which involves making unjust decisions about who lives and who is let (and left) to die—is as dynamic and productive as it is repressive. Modern disasters have been shown to strengthen the power of elites, as states of exception are designed to do, just as they offer opportunities to challenge or resist those very power structures. Moreover, they can forge alternative pathways through alter-political practices that reimagine what engagement can look like in times of crisis (Hage 2015).[15] This engagement might, in fact, be more aligned with the state than oppositional. Consider the question I frequently encountered: "Did people wear masks in Vietnam?" *Yes, they generally did in Hanoi.* Public health measures did not produce a tangible social conflict between civil liberties and public safety. For my Vietnamese interlocutors, many of whom were baffled by American anti-maskers, pandemic policies did not so much violate individual rights as they secured the right to collective life.

Scholars and public intellectuals have observed that when institutions fail in times of crisis, civil society prevails (e.g., Solnit 2009, 305). This argument is supported by examples of inadequate government responses to emergencies, in which communities actively self-organized to ensure their survival and recovery. However, this perspective tends to assume both a weak public sector and a stark divide between state and society rather than a relation of interdependency. It overlooks existing social and political accommodations—those "established norms of cooperation" between state and nonstate actors (Fairhead 2016, 11), which are capable of generating decisive collective responses to a looming catastrophe. Such norms can facilitate the mass mobilization of a largely consenting population promised care and protection, even as they coexist with gestures of refusal.[16]

Instead of adopting a top-down approach centered on docile bodies and repression or a bottom-up perspective emphasizing liberal individualism and resistance, I propose a lateral and relational understanding of a public that was "re-auralized" during the pandemic through sound (Lacey 2013, 30).[17] This attuned public might be better understood in nonbinary terms, as an alliance of cohabiting, co-sounding, and co-listening humans and nonhumans—a collectivity comprising affective bodies, authoritative objects, viral matter, and sonic technologies.[18] While forged through shared risk, pandemic outcomes and vulnerabilities were neither equally shared nor uniformly distributed across this collectivity. A lateral and relational approach, then, allowed me to conceptualize pandemic care work as a form of sonic "meshwork" (Ingold 2011), woven through acoustic practices and co-present listening. This mediated meshwork engaged the senses and evoked mutual concern, binding relational beings and things—coronaviruses included—through acts of care and affects of fear.[19]

<h2 style="text-align:center">SENSORY ATTUNEMENTS: RETHINKING
FIELDWORK AS SOUNDWORK</h2>

The COVID-19 pandemic transformed not only how we live but how we conduct anthropological research, particularly by reshaping our sensory engagement with the world. The crisis challenged established methodological practices while inspiring novel approaches to ethnographic inquiry.[20] At the heart of this disruption lay anthropology's proprietorial relationship to "the field," a concept which continues to define the discipline's core values and identity despite long-standing criticism of its colonial underpinnings.

With field sites rendered inaccessible, researchers found creative ways to adapt. Some championed the merits of digital ethnography, embracing virtual methods as substitutes for physically "being there." Others—particularly those based in the Global North—relied on local research assistants to conduct "proxy" fieldwork, collecting and transmitting data remotely on their behalf.[21] While these methods generated important insights into the pandemic's impact on human and nonhuman life—especially the lives of our research collaborators—they often fell short of capturing the embodied, place-based experiences central to ethnographic research.

Certainly, pandemic regulations, like lockdowns, radically altered the forms of intimacy and modes of participation that ethnographers embedded in their research sites could engage in. Rather than eliminating fieldwork, however, the pandemic provided an opportunity to rethink ethnography along the axes of multimodality and multisensory methodology.[22] These approaches could better account for the ways technologies, in times of crisis, mediated relationships marked by separation—reconfiguring how people sensed, connected, and were governed from a distance.

Liz Przybylski's (2021) concept of hybrid ethnography offers a useful framework for addressing these methodological challenges. The pandemic, more than any other global health emergency, underscored how technological changes transformed communication, dissolving the sensory and spatial boundaries between the "purely digital" and the "purely physical" (2021, 174). For example, it catalyzed shifts to e-government, as states increasingly relied on digital platforms to monitor public health. People's everyday engagement with media technologies and the infrastructures that support them reshaped how and when they accessed information and navigated regulatory frameworks. Online and offline forms of care and connectedness became increasingly difficult to disaggregate as "the field" itself became more hybrid. In Hanoi, I experienced this hybridization through a range of relationships and their associated modalities, from virtual interactions with state-assigned health care workers to spatially distant, in-person encounters mediated by protective technologies like plastic partitions.

To attune myself to the hybridity of these relational dynamics and their shifting forms of "bodily and sensory co-presence" amid social isolation (Low and Abdullah 2021, 54), I turned to embodied methodologies of co-present listening and walking-with, guided by intuitive movement and sensory re-engagement with the urban environment. While much of the world was "closed," Vietnam remained internally "open" (except for a three-week national lockdown), allowing me to move through daily life alongside my neighbors with a sense of care and attentiveness. This afforded a close-up perspective on how the Vietnamese state managed a public health emergency, highlighting the role of sound-based strategies deployed across all levels of government and the ways people responded to these sonic interventions.

These unique circumstances called for an autoethnographic approach that placed care and collective experience at the center of feminist-informed, multisensory research (Guzman and Hong 2022), illustrating its potential to reimagine ethnographic practice in a hybridized sensory world. My use of autoethnography is not intended as a retrospective analysis of my personal experiences as a white cisgender woman to illuminate broader cultural patterns. Rather, I take my cue from Elizabeth Chin (2016, 189, 196), who insists that autoethnographers are, first and foremost, *ethnographers*. This suggests that we are not merely thoughtful observers or skilled narrators of social life but engaged practitioners—and, crucially, attentive listeners—actively and intentionally manifesting a more caring, creative, and reflexive anthropology. Guided by a feminist ethos of care, this approach encourages deeper reflection on power and inequality, including our own world-making knowledge practices (Staffa, Riechers, and Martín-López 2022), while attending to "matters of care" in how we observe, represent, and interconnect with both human and nonhuman worlds (Puig de la Bellacasa 2011, 99–100).

In the absence of group dynamics, autoethnography allowed me to foreground my own sensibilities and sense-making of the pandemic while sensing alongside others, and doing so in careful compliance with COVID-19 mandates. Tuning in to the urban rhythms of sounds, noises, and murmurs with an "attentive ear" (Lefebvre 2013, 27) provided unique sensory experiences of place during isolation, enabling deeper intersubjective moments of attuned "with-ness," despite the separations imposed by layered "architectures of protection" (Mattern 2020).

Adhering to an ethics of care, this book makes no claims to knowledge beyond my own perception and situated sensory interactions with the urban environment, nor does it attempt to speak for how others listened, heard, or made sense of sound.[23] Senses and sensations—"socially made and mediated" through race, gender, culture and class—are neither perceived nor interpreted uniformly but shaped by power and positionality (Hsu 2008, 433). Because the senses transmit cultural values (Classen 1997), including those that support evolutionary teleologies and white supremacist hierarchies, they often become subject to moral and disciplinary regulation or state control (Howes 2019, 22).[24] Scholars exploring sound and Blackness in the United States, for instance, show how Black sensory perception has been especially targeted—constituting a racialized matter of life and death (Stoever 2016), while also being reclaimed as a powerful tool of dissent (Brooks 2006). This tension between domination and subversion makes the senses particularly good to think with (see Low 2023), above all in the context of a public health emergency.

Taking care to cultivate relations that would not contribute to viral transmission, I approached this sensory terrain not through conventional fieldwork but through *soundwork*. This methodology prioritized the listening "ethnographic ear" over the observing ethnographic eye (Erlmann 2004; Ochoa Gautier 2014), emphasizing situated *participant listening* rather than participant observation (Pink 2015). By favoring earwitness over eyewitness accounts (yamomo 2018, 49), my approach foregrounded sound as a critical sensory register for relational ethnographic inquiry.

This auditory engagement was particularly fitting given the sonic geographies that emerged in response to COVID-19, when visual contact was limited due to isolation policies. My initial encounters with Vietnamese pandemic governance, for instance, were mediated through digital sounds and technologies, as I elaborate in act 1, and these would continue to serve as key media infrastructures for care and coordination. Indeed, the novel coronavirus was both detected and monitored through a range of sonic cues, from symptomatic coughing to the beeps and alarms of diagnostic devices.[25] Under pandemic conditions, acoustic atmospheres emerged as both the object and method of autoethnography, highlighting the relevance of soundwork for documenting altered sensory worlds.[26]

WHAT DO SOUNDS WANT? ON THE SONIC GOVERNANCE OF CARE

This book invites a shift from seeing to listening, offering sound as an entry point for analyzing regulatory institutions. Scholars—including myself in earlier works (Schwenkel 2009a; 2020)—have made extensive use of the concept of "the gaze" in visual domains to examine relations of knowledge, power and discipline. However, Lauri Siisiäinen (2013) notes in *Foucault and the Politics of Hearing* that

Foucault himself challenged ocularcentrism by highlighting the importance of sound in nineteenth-century clinical medicine—considering it an integral part of the "sensory triangulation" of the "sight/touch/hearing trinity" (Foucault 1973, 163–4), despite declaring an eventual "triumph of the gaze" (Siisiäinen 2013, 165). Foucault's ideas about sound and hearing as entwined with disciplinary power are useful for thinking about the governance of care during the pandemic, not to advocate for a reordering of the senses but to recognize the role of sounds in shaping medical knowledge production and emergent care practices.

"Care," in this context, encompasses the collective work done to ensure a livable, sustainable, and interconnected flourishing of human and nonhuman life and environments (Care Collective 2020; Tronto 2013). Yet care also extends beyond positive affects, good deeds, or kind gestures (Murphy 2015, 719). As a relation of power, care often excludes those deemed unworthy of being cared for (Puig de la Bellacasa 2011, 95). Who or what is deserving of care, and how that care is expressed and distributed, reflects the dominant values upheld by society.

In Vietnam, these values manifest through what Susan Bayly (2020, 1587) describes as an "interactive sensibility"—an affective mode of relational care grounded in an ethics of mutuality. Aligned with Annemarie Mol's (2008) "logics of care," health care is understood to be responsive, attuned to bodies-in-relation and embedded in broader cultural systems and networks of social obligation, rather than in individual choice or action alone. Examining the sonic dimensions of care governance during a state of emergency provides a framework for understanding the contradictions that may emerge from these care logics and relational sensibilities. Through sound, the exercise of power rendered the boundary between care and control unstable, revealing a paradox within this "interactive sensibility"—how practices intended as care may be experienced as coercive or even harmful (Murphy 2015, 720).

In his study of Cuba's socialized health care system, Sean Brotherton observes that state biopower operates through a constellation of adaptive strategies rather than as a static monolith (2012, 8). This responsive approach to public health governance was evident in pandemic-era Hanoi, where sound emerged as a key disciplinary technology to promote collective well-being in alignment with socialist ideals. Sounds acted as catalysts, urging people into action and inviting them into imagined collectivities of care through attentive listening and the formation of a "public ear" (Peterson 2021). This dynamic reflects Jane Tronto's observations on the relationship between citizenship and caring. While care can offer essential support, it also carries expectations of civic responsibility. Tronto highlights this dual nature, noting that care can become a "burden," such as the obligation to "maintain and preserve political institutions" as a condition for membership in wider society (2013, x). In the context of Vietnam, sonic strategies of governance helped translate care into a broader civil project—one that framed mutual support, attentive listening, and collective caretaking as essential duties of socialist citizenship.

Sonic governance shaped but did not determine people's actions, however. Individuals were not merely passive subjects of state regulation or sonic manipulation but active participants in the co-creation of acoustic environments. People engaged creatively with sound, acting as sonic agents (Labelle 2018) who reconfigured state-produced soundscapes through acts of appropriation and reinterpretation. This engagement often challenged state claims to sonic authority while imbuing public health messages with new meaning, as exemplified by the global response to the Ministry of Health's video "Ghen Cô Vy" (see preface). Consequently, people were not merely acted upon by sounds during the pandemic but actively shaped sonic practices and the forms of care they transmitted.

Some readers may question my focus on the acoustic qualities of pandemic governance, finding it difficult to reconcile with more familiar panoptic frameworks of disciplinary power. Many observers have assumed that visual representations, such as posters and billboards, were the primary means of Vietnamese state persuasion and socialist mobilization against COVID-19. Indeed, literature on Vietnam's pandemic response written from afar often overemphasized state socialist imagery, creating an impression that it dominated state communication.[27] This emphasis on visual media is understandable given the strong theoretical focus on the aesthetic dimensions of governance in scholarship, much of which is inspired by Foucault's work on surveillance. But effectively governing remotely during COVID-19 necessitated a (re)turn to aurality. While pandemic iconography was not entirely absent, its visibility and relevance were limited in daily life, with people confined indoors, making sonic interventions a more immediate and pervasive expression of state presence.

By contrast, sonic technologies that communicated concern for the population's health and welfare proved more dynamic and panaudic in their scope. With their ability to penetrate walls and bridge distances, sounds were arguably more immediate and effective in reaching listening subjects, shaping their everyday sensibilities in ways that visual media could not match. My analysis does not suggest the totalizing power of sound but rather its privileging at a particular crisis moment (Sterne 2003, 4). Sonic dominance did not imply the erasure of the visual—given the interdependence of sight and sound—but rather its relegation to a less prominent sensory register. Without establishing false binaries between seeing and hearing, this approach attends to the distinct insights and outcomes that emerge when we ask not what pandemics *look* like but what they *sound* like, and how their effects and echoes resound.

Like images (Mitchell 2004), we might think of sounds as animated forces that have the capacity to desire, effect, and produce affects. Under a declared state of emergency, sounds drew citizens into relations of care and discipline, grounded—at least initially—in willful acts of collaboration rather than top-down impositions of power (Trnka 2020, 368). While promoting a stern but optimistic vision of social cohesion that allowed little space for alternative perspectives, sounds alone could not quell dissent or prevent noncompliance. During the pandemic, state-produced sounds aspired to many objectives—to amplify authority, foster trust in science while combating superstition, strengthen the legitimacy of party rule, promote collective responsibility, and ensure a healthy population fully committed to mutual care—all while advancing the principles of socialism. As enablers of regulatory processes, sounds were essential to the movement of ideas, resources, technologies, and the very caretakers of populations facing an unpredictable and destabilizing crisis.

Understanding Vietnam's pandemic response in 2020 requires setting aside sight-based assumptions to consider the non-visual techniques deployed to manage a global health emergency. This means taking seriously the state's exercise of power through sound to advance biopolitical goals of protecting collective life and sustaining socialism through the prevention of disease. Far from a novel development, these governance strategies drew on a longer history of sonic mobilization for socialist state formation and the promotion of hygienic citizenship.

SONIC SOCIALISM: ACOUSTIC HISTORIES
OF PUBLIC HYGIENE IN VIETNAM

Vietnam's history of public health regulation is deeply entwined with the country's evolving sonic cultures. From colonialism to the present, sound has served as a powerful vehicle for monitoring, educating, and governing the population in pursuit of disease prevention. The country's rich oral traditions and musical expressions have long played a central role in mobilizing society, disseminating medical knowledge, and imparting social values, with successive regimes adapting these cultural forms to their own ends. The history of hygiene education, in particular, can be traced through song and spoken poetry, which emphasized the importance of bodily self-management as a civic responsibility to curb the spread of infectious diseases.

Sound was deployed as a mechanism of empire to advance its civilizing mission (yamomo 2018). Under French rule in Vietnam, the urban governance of disease gave rise to a distinct soundscape and sonic vocabulary that tied assessments of health and morality to markers of progress and modernity. Disease became a matter of social order and security, with auditory practices serving as sites of intervention, techniques of bodily supervision, and instruments for moral reform. Vũ Trọng Phụng (2011) underscores the gender and racial dimensions of these practices in his satirical reportage, *Lục Xì*, a story about sex workers and venereal disease in colonial Hanoi. The commercial sex industry in Vietnam had long-standing and well-known connections to *ả đào* or singing establishments located on the outskirts of the city. French authorities regarded these venues as musical houses of prostitution that operated as source points for the spread of sexually transmitted infections (2011, 52).

Such framings reveal how the gendered and racialized associations between sound, sex, and disease became embedded in the colonial regime of surveillance. In Hanoi, the French Service des Mœurs, or morality agents charged with monitoring prostitution, were referred to in Vietnamese as the Cảnh sát Xướng kỹ, meaning the police of women who sang for money (Malarney 2011, 14). Infected sex workers confined to the state medical dispensary were subjected to hygiene lessons that involved collective singing and recitation as part of their "cure." While the gynecological examination relied on visual inspection (thus the expression *Lục Xì* or "look-see" for the dispensary), the treatment of the "hygiene problem" embraced auditory methods. Given the high rate of illiteracy among colonial subjects at that time, health education incorporated the performance of musical verse to underscore the importance of cleanliness for the greater public good.[28] Colonial biopolitics, through sonic instruction and a strict daily regimen, ostensibly sought not to punish sexual transgressors but to protect their lives and well-being (Vũ 2011, 100)—and, by extension, the health of the white male population partaking in paid sexual services.[29] These European men were arguably the intended beneficiaries of the hygiene training imposed on Vietnamese women.

Sonic hygiene practices were integral not only to the expansion of colonial biopower but also to the project of socialist nation-building. In both instances, auditory methods utilized health infrastructures to achieve their respective goals. While colonial health authorities used song to instruct and "civilize" wayward bodies in *urban* spaces, the revolutionary state similarly harnessed oral traditions to mobilize and educate *rural* populations. Sound was intrinsic to decolonization and the formation of a sovereign Vietnamese nation-state, where strong, healthy bodies symbolized the emergence of "new socialist persons" (*con người mới xã hội chủ nghĩa*). In this context, sonic practices of care were propagated not as instruments of subjugation, but as a pathway to emancipation—freedom from colonial disease and backwardness (Malarney 2012, 111), ultimately fostering a spirit of hygiene nationalism.

During the revolution, musical forms were repurposed "in the service of modern epidemiology" (Marr 1981, 187). Traveling medical teams harnessed aural cultural practices, using singing to teach germ theory through both direct instruction and folk song primers (Malarney 2012) Singing and personal hygiene were regarded as embodiments of civilization, with cleanliness standing as a testament to love of one's country, as exemplified in the popular motto "vệ sinh là yêu nước," or "hygiene is patriotic" (Craig 2002, 56; see also Aso 2013). This dictum later underwent modification during the pandemic (see act 2).

Decolonization also spurred the rapid development of Vietnamese media technologies within an emerging socialist sonic infrastructure. These included Vietnamese-language radio broadcasting and public address systems deployed for anticolonial mobilization and mass communication (Ó Briain 2022). A dynamic wartime soundscape took shape, characterized by alerts and alarms designed to strengthen conflict readiness. This network of auditory cues and prompts enhanced both defense capabilities and national unity, as collective sound-making and vigilant listening helped protect vulnerable lives from external threats. Sonic solidarity forged through anti-imperial struggle would come to anchor Vietnam's postcolonial identity.

The historical relationship between sound and regimes of care has received little sustained attention in recent ethnographic research on Vietnam. This gap reflects the limited intersection to date between sound studies and medical anthropology, despite growing interest in both fields, with Tom Rice's (2013) *Hearing and the Hospital* standing out as a notable exception. While scholarship on Vietnamese caregiving and therapeutic practice has examined evolving concepts and forms of care as both resource and contingent relational practice (Tran 2023, 44; Shohet 2021), their sensory dimensions, particularly communal contexts of medical sounding and listening, remain underexplored. Notably, the role of *visual* biotechnologies in health care decision-making has been well documented (Gammeltoft 2014). Several important studies have also focused on the governance of care, especially the control and prevention of infectious disease (Lincoln 2021; Porter 2019). I draw inspiration from these seminal works to foreground the intersection of sonic politics and biopolitics, using sound as a lens to examine both the relational, affective contours of care and the regulatory functions of the state during disease outbreaks, including COVID-19—a dynamic I call the sonic governance of care.

Sonic governance of care did not emerge in isolation; rather, it manifested as a strategic tool within a broader framework of disaster response. Specifically, it became a critical component of what might be called "disaster socialism," a concept describing the centralized management of national emergencies that includes targeted acoustic interventions to uphold core socialist values. I propose this term—and its sonic socialist components—as a counterpoint to a profit-driven model of "disaster capitalism" (Klein 2007), which prioritizes the health of the economy over the welfare of the population in disaster mitigation. In contrast, disaster socialism draws on a robust public sphere and state investment in emergency infrastructure—advancing planning, preparedness, coordination, and resilience—to administer life and dispense care collectively without resorting to market-based solutions, even if it does not always do so equitably. This framework underscores the endurance of deeply embedded logics of social mobilization, amplified through sonic infrastructures that were (re)activated during the pandemic to uphold a socialist ethos of care. Public health campaigns extended beyond mere hygiene instruction—long a staple of socialist messaging—serving as acoustic media for shaping individuals into healthy and productive socialist citizens.

A sonic approach to public health challenges the notion that socialist ideals—including those that that once championed universal health care—have simply eroded in the post-Soviet era. Instead, aligning with scholars of "revolutionary medicine," it views the public sector as a fertile ground for examining the contradictory operations of state biopower in the aftermath of socialist restructuring (Brotherton 2012, 184). From this perspective, policies promoting self-management warrant critical scrutiny, not as affirmations of autonomous neoliberal subjects, but as expressions of interdependence within a broader collectivity, one that remains a cornerstone of social well-being (Schwenkel and Leshkowich 2012). This standpoint seeks to understand the dynamics of Vietnamese crisis management on its own terms rather than through inherited frameworks of neoliberalism. While Vietnam has embraced global capitalism through its "socialist-oriented market economy" (*kinh tế thị trường định hướng xã hội chủ nghĩa*), it continues to govern under a one-party system whose discourse frames loyalty and social duty as moral obligations to prioritize national interests over individual advancement.[30] Socialist biopower, particularly in its governance of public health, operates by instilling a shared sense of collective responsibility and mutual concern for the common good, rather than emphasizing personal rights and freedoms (Gammeltoft 2007, 160), an ethos that became especially salient during COVID-19.

Facing a global health crisis, the Vietnamese state activated a sound-based emergency response rooted in its disaster socialist framework of social mobilization. Much as they have historically, sonic technologies enabled rapid risk communication and mass dissemination of epidemiological information, forming the basis of its state-managed care. Crisis media infrastructure—developed during the US war and repurposed for the pandemic—merged "residual" media (Acland 2007) with new digital practices, often re-sonifying environments through the revival of "retired" technologies that evoked auditory memories of past catastrophes.[31] Authorities deployed these sound-based interventions as "sonic instruments of biopower" to discipline at-risk bodies (Malmström 2021, 604). This "sonic socialism," as I conceptualize these practices, reveals how a landscape of media technologies and measures of biosecurity forged in crisis has reshaped the relationship between power and care.

SONIC MAPPING: TRACING SOUNDWAVES ACROSS PANDEMIC TIME

This book unfolds as a sensory autoethnography of the first year of COVID-19, tracing my movements and modes of listening under evolving conditions of pandemic governance. To capture shifting orientations across disquieting experiences of isolation, I organize "pandemic time" into three acts—quarantine, lockdown, and spatial distancing. Each act centers on a specific containment policy, along with its attendant sonic practices and underlying logic of care. With crisis managed remotely, people primarily engaged with the state through its mediated technologies, especially sound-based ones, designed to cultivate an imagined listening public bound by shared commitments to public health.

I use the term "acts" rather than "chapters" to illustrate the performative and episodic qualities of pandemic life. Collectively, they chart changing contours of care—from state care to self-care and ultimately mutual care, while raising critical questions about who cares and for whom. The narrative moves chronologically across the first ten months of COVID-19, beginning with Patient 17's neighborhood outbreak on March 6, continuing through the first recorded coronavirus-related death on July 31, and extending into the uncertain months that followed. Yet, as the acts reveal, pandemic temporality was neither linear nor uniform but manifested as fragmented and uneven. This rhythm is mirrored in the narrative structure of each act, which traverses time and space to evoke the lived—and sensed—disruption of everyday life.

Act 1 chronicles my arrival in Vietnam and initial encounters with the state's emergency preparedness apparatus, mediated through digital technologies and e-government platforms. This set the stage for attuning my ethnographic ear to the sonic governance of care unfolding in Hanoi. Quarantine became the context for analyzing the sensory effects and affects of text messages and notifications from health authorities, who monitored my well-being through real-time communication. By examining the ambiguity between state care and control embedded in digital health infrastructures, I reveal how medical supervision rendered the state more personal and familiar. This reliance on remote surveillance through virtual interaction introduced an intimate form of crisis management, one that structured the rhythms of my social isolation through the routine sounds of ringtones and alerts.

As pandemic conditions evolved and governance adapted in response, so too did the contexts for collective sounding and listening. Act 2 explores these changes during the national lockdown that followed my quarantine release. Focusing on the distinctive temporalities of mass communication technologies, it traces shifts in my own sonic sensibility from anticipatory listening *for* sonic events, like official text alerts, to co-present listening *with* others during public broadcasts. This listening reorientation reflected a broader transition from individualized state care to collective self-care mediated through public health messaging. The pandemic revival of loudspeakers, the focus of this act, sparked intergenerational controversy. Integral to Hanoi's sonic history, these technologies served as sites where power was both exercised and contested through sound. While state authority was projected through sonic dominance over urban space, instances of cooperation coexisted with acts of sonic dissent.

Act 3 marks a spatial and methodological shift, from "participant listening" within domestic spaces of the home to "walking with" in outdoor public spaces through sensory strolls. Mandatory spatial distancing introduced a new urban normal, altering how people perceived, used, moved through, and "sounded" in public space. As an atmosphere of normalcy gradually returned, humans and nonhumans reasserted their sonic belonging, creating a more diversified acoustic ecology that prompted shifts in my ambulant soundwork practice. Pandemic soundwalking fostered renewed sensory connections between people and nature in the city as familiar sounds—pet bird calls, street vendor cries—slowly reemerged. Distanced "strolling alongside" enacted a shared ethics of care and a commitment to protecting communal life, even as the state's sonic presence continued to resonate, reminding listeners of the need to remain care-full and vigilant.

To illustrate the human impact of pandemic inequalities, I present five iconic "social types" or symbolic groups of non-state actors I encountered through everyday sounds.[32] These pandemic figures, as I call them, played a critical but often under-analyzed role in the sonic work of care during this exceptional period. Positioned between acts, these figures both intersect with and disrupt the flow of the text, much like the pandemic intersected with and disrupted lives. Unlike the mostly anonymous individuals in the three acts, where social isolation shaped a sensory autoethnographic mode of storytelling, these figures offer more conventional ethnographic close-ups, with identified interlocutors. Through these portraits, I demonstrate how pandemic uncertainties manifested in the daily lives of overlooked and unheard individuals.

These figures are superspreaders (Patient 17), essential workers (the security guard), domestic workers (the housekeeper), nonessential workers (the Grab driver), and migrant workers (the Philippine singer). With the exception of the superspreader who compromised care, all were urban care workers who endured precarious conditions to perform their low-paid jobs—work centered on regulating and reproducing social life. Together, their experiences exemplify how pandemic biopolitics functioned as a politics of differential vulnerability (Lorenzini 2021), heightening their uneven exposure to risk.

A brief coda examines how the Delta variant dramatically transformed the conditions for sonic socialism during the second year of the pandemic. Vietnam's centralized model of "disaster socialism" evolved in response to surging infections and deaths, revealing tensions between anticipatory planning and adaptive governance. Rap music exemplifies how public health institutions balanced control with cultural innovation to appeal to wider audiences, particularly youth who might otherwise tune out official messaging. Musicians not only amplified public health directives but filled gaps in state care through mutual aid initiatives, demonstrating how non-state actors, in this case hip hop artists, were incorporated into state biopolitics rather than silenced by sonic governance.

Essential Workers

The Security Guard (Người Bảo Vệ)

Disasters expose and amplify precarity, even as the precariat may assume greater public sympathy in times of national emergency (Allison 2013, 8). Consequently, the onset of COVID-19 exacerbated precarious living and working conditions, resulting in the uneven distribution of vulnerability that scholars have identified as "pandemic precarity" (Perry, Aronson, and Pescosolido 2021). This was especially the case with those laboring in the informal economy. Replaceable *essential* workers, often migrants with no social protections whose job was to safeguard others, shouldered a disproportionate risk of exposure to COVID-19 (Cassiman, Eriksen, and Meinert 2022). In Vietnam, the figure of the security guard, or *người bảo vệ*—literally translated as the person (*người*) who protects or defends (*bảo vệ*)—epitomizes this low-wage, high-risk workforce, which served at the pandemic's frontlines through direct and consistent contact with a potentially infectious public.

The người bảo vệ is ubiquitous in Vietnam. Hired to monitor entryways to banks, stores, restaurants, offices, government buildings, apartment complexes, and other business or service establishments, they straddle formal and informal economies, marking the threshold between public and private domains. Unarmed and only in uniform in larger operations or government institutions, they typically fall outside the prime working age of twenty-five to fifty-four and are almost always men. For senior guards, inadequate pensions have compelled their exit from mandatory retirement to seek unregulated sources of income, signaling the precarity of aging as a growing social phenomenon in Vietnam. This was the case with Hùng, a người bảo vệ who was hired by my landlord to watch over her ten-story house that she converted into lucrative apartment rentals.[1] After retiring as a hydropower technician from the state electric company in a neighboring province,

Hùng became a security guard in my building to help pay for his ill wife's mounting health care costs. He earned an additional six million VND (US$250) monthly by working twenty-four-hour shifts on a week-on, week-off rotation until the pandemic disrupted this schedule. The socialist state had failed to fulfill its promise of care and protection in Hùng's older years, despite his years of labor and loyalty to nation-building. This situation also echoed that of the second guard, Văn, a retired war veteran from outside Hanoi. Without adequate social protections, he had little choice but to return to "defense" work—this time guarding the property of the elite rather than the territory of the state.

Far from being unskilled laborers, senior security guards in particular embodied versatility, drawing on a range of untapped competencies from their lived experiences and training, including from wartime. Some had attended technical schools in the Eastern Bloc but faced unemployment or job insecurity after the Soviet Union's collapse (Schwenkel 2022a). Their expertise enabled them to take on tasks and roles that extended well beyond basic guarding duties. Người bảo vệ spend much of their time either sitting or standing around while keeping watch, serving as the "eyes on the street" (Jacobs 1961) who help maintain social order. For instance, they manage the flow of traffic, direct parking, and guide pedestrians. Like door attendants or bouncers, they control access to buildings, wielding the authority to permit or deny entry. Many also perform maintenance tasks or provide supplementary services. Hùng, for example, had a knack for repairing broken appliances, which generated extra income on the side. During slow periods, người bảo vệ can be found socializing at nearby corners or tea stalls while maintaining a line of sight to their workplace. Like other migrant workers living precariously in the city, they are essential to Hanoi's informal caretaking economy yet remain easily expendable, further heightening their vulnerability.

While heroic frontline workers—especially the almost superhuman health care professionals, who in Vietnam were honored as "*chiến sĩ áo trắng*," or soldiers in white dress—have received extensive media attention and celebration, other essential workers have remained largely overlooked. During the pandemic, the state deemed certain in-person work necessary for maintaining the "provision of essential goods and services" (*cung ứng hàng hóa dịch vụ thiết yếu*) to ensure economic resilience, yet these workers garnered little recognition. Even more invisible were precarious workers like security guards, who constituted a crucial part of the urban "essential workforce" (*lực lượng lao động thiết yếu*) in establishments exempted from closure mandates. By "essential," I refer not only to how người bảo vệ contributed to the city's frontline response through their public-facing interactions with service-seeking persons but also to their expanded monitoring responsibilities. Following the onset of COVID-19, these responsibilities grew to include health care related tasks, including the refinement of what we might term "techniques of medical listening" to ensure the safety of others.[2]

The arrival of the pandemic thrust security guards into new roles as front-line health screeners, broadening their traditional duties of visual surveillance and property caretaking to auditory monitoring and safeguarding of public well-being. Attuning themselves to symptomatic cues—such as coughing—and sound-emitting technologies, these workers became crucial decoders of the nuanced boundary between health and disease. With minimal training, they were tasked with screening for indicators of COVID-19 while using infrared thermometers, whose beeps, readings, and alerts created a new sensory landscape of medical surveillance. Reminiscent of the stethoscope's diagnostic power (Rice 2013; Sterne 2003, 91), these devices transformed guards into medically attuned listeners and interpreters of signs.

The temperature screening procedure quickly became routinized and universalized: halt one meter before the guard, who, equipped with their instrument of biopower, extended their arm to aim the sensor at one's forehead until it emitted a sharp "beep." Afterward, a cautious nod followed their analysis of the digital reading (figure 3). Beyond non-contact body checks to diagnose health status, người bảo vệ were tasked with enforcing public compliance with safety protocols, including mask mandates and compulsory hand sanitization before entering premises. On the frontline of infectious disease prevention, they labored under hazardous conditions with increased risk of contagion to safeguard those on the inside from the viral dangers circulating outside. The ubiquitous beeping of infrared devices echoed their duty to protect both individual well-being and the broader health of the nation.

Người bảo vệ, though proficient in standardized monitoring procedures, lacked extensive training for responding to situations of unanticipated risk. Their medical listening, therefore, relied heavily on intuitive knowledge to interpret threats of contagion. The routinization of screening, intended as a form of bodily discipline to manage public health vulnerabilities, hinged on presumptions of stability and predictability—normative respiratory sounds and body temperatures indicating an absence of disease—to maintain a sense of biosecurity. Deviations, as exceptional events, disrupted the system's rhythm, throwing it into momentary chaos. Guards, in other words, found themselves ill-prepared to handle glitches and anomalies when alerts signaled a potential breach of public safety.

I experienced this unpreparedness firsthand on one occasion when my temperature registered at 35.7 degrees Celsius (96.3 degrees Fahrenheit), eliciting a perplexed response. After the thermometer sounded, the guard, a young man with a friendly demeanor, looked at me with confusion and showed me the number. He then called over an older colleague, who also studied the reading before turning to me with bewilderment. Both guards had been trained to listen for the affirmative beep and to watch for high temperatures, but they appeared to lack clear protocols for interpreting unusually low numbers. What could it mean? Was it also symptomatic of COVID-19? I smiled and joked about the frigid blast of

FIGURE 3. Temperature screening with acoustic feedback to monitor COVID-19 in a Hanoi office building, 2020. Photo by the author.

air conditioning, mentioning how it prompted some Vietnamese women to wear winter coats inside buildings. This explanation resonated with the guards. They laughed, easing the tension, and waved me into the store.

Security guards were inclined to attribute anomalous readings to environmental factors, such as hot or cool air, rather than viewing them as indicators of illness. The fact that elevated temperatures did not always trigger the anticipated response (denial of entry) suggested that thermal discipline was designed around expectations of self-regulation rather than strict securitization. That is, it was assumed that someone feverish would make the socially responsible choice to self-quarantine rather than venture out in public. When faced with ambiguous alerts that did not clearly indicate health, người bảo vệ relied on their own observations and experience to make sense through interpretive techniques of medical listening. For example, they advised individuals with high temperatures but no cough to seek relief from the outdoor heat in the *bóng râm* (shade) before retesting, externalizing risk to atmospheric conditions. In this way, the screening procedure itself might have functioned as little more than an optics—or an acoustics—of public reassurance, offering a false sense of security to those whose vetted bodies passed the medical check.

I contemplated this one muggy summer evening in 2020, during spatial distancing (see act 3), while out for a drink with my spouse. As we were the sole customers at a neighborhood pub, the manager sat down to discuss the pandemic's impact on his business, which had reopened just a few weeks earlier after the April lockdown (see act 2). The door swung open, and a sizable group of Vietnamese patrons, both men and women, entered. The manager quickly stood to greet them, while the cashier offered hand sanitizer to each person and the security guard readied the thermometer to scan their temperatures. After each beep, the customer proceeded upstairs to the second floor. However, when the last person, a woman, registered what sounded to me like a typical beep, the security guard turned to show the manager an irregular reading. A panicked look crossed the manager's face. "Stop her!" he demanded as the woman darted up the stairs. "Chị ơi!" (Hey big sister!), the người bảo vệ yelled, chasing after her. She paused for a second temperature check. The device beeped again. Her male companion, who had also stopped, gave the much younger guard a stern look as the guard stepped aside to let them pass. "How high is it?" I inquired, watching the guard slowly descend after she disappeared above. He held up the digital display for me to read: 37.4 degrees Celsius or 99.3 degrees Fahrenheit.

No one seemed to know how to proceed. The guard's role was to detect deviations, not necessarily to intervene, leaving the employees suspended in a tense state of uncertainty. "It's high, isn't it?" the manager asked, his voice tinged with worry. I shrugged. "What's the threshold?" His jaw clenched. "The cutoff point is 37.5 degrees," he replied, clearly weighing the delicate balance between health risks and financial reward. The young guard waited fifteen minutes before returning upstairs to recheck the woman's temperature, allowing her time to recover from the outdoor heat, I was told. Ultimately, the female guest was permitted to stay, but an undercurrent of anxiety lingered as the possibility of viral infection still hung in the air. Though security guards possessed both the authority and the diagnostic technologies to monitor bodies, including through medical listening, their actual power to control those deemed public health risks was surprisingly limited in practice.

Act 1

Quarantine

State Care

A familiar sound marked my first encounter with COVID-19 in Vietnam: the chime of an incoming text message alert. The mostly empty plane from Taipei had touched down in Hanoi just after noon, thirty minutes behind schedule, on a cool and overcast day on March 17, 2020. As we taxied on the runway, I powered on my phone, its US-based SIM card connecting to the local network. Almost immediately, an English-language notification flashed on the screen, signaling the start of uncertain times ahead:

VIETNAM_MOH

March 17, 2020, 12:02

Vietnam Ministry of Health: welcome to Vietnam! If you've travelled from an area affected by COVID-19, you are requested to complete a medical declaration form at the designated booths before the immigration counters [figure 4]. For COVID-19-related advice while in Viet Nam, please call the hotline 19009095 or visit ncov.moh.gov.vn. You may also download the SỨC KHỎE VIỆT NAM app for more information. Stay healthy and do your part in preventing and controlling the spread of COVID-19. Wishing you an enjoyable time in Vietnam.

A minute later, I activated my second phone, equipped with a local Viettel SIM card. It too pinged and buzzed, this time with a similar instant message in Vietnamese:

BO_Y_TE

March 17, 2020, 12:04

Chào mừng quý khách tới Việt Nam! Nếu quý khách đi qua khu vực có dịch COVID-19, hãy khai báo y tế trước khi nhập cảnh để được trợ giúp. Hãy gọi

19009095 hoặc truy cập website ncov.moh.gov.vn để được tư vấn. Hãy tải ứng dụng SỨC KHỎE VIỆT NAM để tương tác theo dõi sức khỏe trong thời gian ở Việt Nam. Hãy có trách nhiệm với bản thân, gia đình và cộng đồng để chung tay ngăn chặn dịch bệnh COVID-19. Chúc quý khách mạnh khỏe!

Despite their similar content, these two digital messages from the central government employed distinct discursive strategies of "soft authority" while advancing a comparable politics of civility. Both utilized courteous, noncoercive language to frame compliance as responsible and voluntary. The English version "requested" arrivals to follow standard health procedures, inviting them to proactively acquire information about the virus through multiple communication platforms (telephone, app, website). In contrast, the Vietnamese text repeatedly used the polite imperative marker "hãy," which softened the tone of directives while encouraging individual action for the public good. More broadly, the messages differed in their conceptualization of responsibility. The English-language text advocated for individual accountability, urging visitors to stay informed and be mindful of not spreading the virus. Conversely, the Vietnamese version framed responsibility as *relational*, emphasizing the need to "be responsible for oneself, family and community" (*hãy có trách nhiệm với bản thân, gia đình và cộng đồng*), while calling on arrivals to "join hands" (*chung tay*) in stopping the virus.

This chapter examines how digital technologies constituted sensory environments during COVID-19, particularly through sound-based media that shaped perceptions and embedded people in wider sonic collectivities. As the text messages revealed, the pandemic unfolded as a digitally mediated experience, with the state leveraging technological affordances to foster a shared sense of purpose, mutual care, and solidarity between citizens and authorities. A vast media infrastructure, heavily reliant on sound but engaging multiple senses, worked to both overcome isolation and control the flow of information, while monitoring and managing the threat of disease.

Among these media technologies, the smartphone stands out as a device designed for multisensory engagement with vision, touch, and sound, shaping daily sensate experiences through ringtones, alerts, music, and video playback. Like in Jamaica (see Horst and Miller 2006, 64), the ubiquity of smartphones in Vietnam has transformed the acoustic environment, with loud conversations, audible streaming, and amplified notifications becoming distinctive features of urban soundscapes. Audio features played a particularly critical role in Vietnam's COVID-19 response, enabling new forms of state care, as exemplified by the viral public health video "Ghen Cô Vy," which bridged visual and sonic elements to engage an attentive public (see preface). This everyday technology served as a sonic medium for mobilizing a largely cooperative population around a shared national objective: to contain the virus and prevent the country from becoming an epicenter of COVID-19. During the public health emergency, smartphones

17/3/2020 In phiếu

Passenger uses this part for entry/exit/transit clearance and for protection of your health

Full name (BLOCK LETTERS): **schwenkel christina**

Province/City/Territory/Country of departure: Hoa Kỳ, California

VERIFICATION BY HEALTH QUARANTINE OFFICER

ĐÃ KIỂM TRA

17-03-2020

Date Month Year 2020

For your own heath and that of the community, if you experience any of the above-mentioned symptoms, please contact heath quarantine units at points entry or the nearest healthcare centre or email to Email: .... or Fax: ....

Hotline of province/city of point of entry: ....

Hotline of the Ministry of Health: 19009095

Declaration code: TK-000119742

Website: https://tokhaiyte.vn

(QRCode code used to Check in, Check out)

https://tokhaiyte.vn/?page=Passenger.Print.detail&groupId=5e5377a276801b3e7607818b&id=5... 1/1

FIGURE 4. Institutional preparedness: mandatory health e-declaration for entering Vietnam, March 2020. Author's collection.

facilitated this goal by rendering the state more intimate and immediate through persistent alerts and vibrations that demanded urgent attention—exemplified by the welcome texts sent by the Ministry of Health.

My use of terms like "cooperation" and "compliance" is intended to complicate oversimplified narratives about power dynamics. These narratives reinforce a flawed democracy-authoritarian binary that equates certain freedoms as inherently "Western"—particularly American—while portraying people in "the East" (especially the "communist East") as docile subjects of an unaccountable state. This Orientalist perspective, grounded in colonial discourse, framed interpretations of responses to COVID-19 across liberal and non-liberal regimes, obscuring more nuanced considerations of how different societies navigated and recalibrated the balance among individual liberties, economic imperatives, and national biosecurity. For some observers, Cold War tropes of communist repression (loss of personal freedom) and deception (false numbers) became the default explanation for Vietnam's low COVID-19 caseloads in 2020. Consequently, measures perceived as restricting individual autonomy, such as mask mandates and collective cohort quarantines, were hastily condemned as hallmarks of totalitarianism, irrespective of their public health merits. Notably, this critique was not applied equally to all Asian countries with low infection rates. Taiwan and Hong Kong, which enacted similarly strict (if not stricter) pandemic measures, largely escaped such international scrutiny.

While caution is warranted regarding how emergency conditions might enable unchecked expansion and centralization of state power (Agamben 2005), pandemic governance in Vietnam—particularly its sonic governance of care—unfolded in more complex and multifaceted ways than is often assumed.[1] Rather than relying on overt domination or coercion, it operated through more persuasive and interactive strategies, including digital technologies to regulate human conduct, such as e-government services and immersive e-care initiatives. The instant messages and mobile alerts I received upon my arrival in the country—which people could silence or turn off—illustrate this approach. This aligns with Foucault's argument that power is neither inherently violent nor based on implicit consent. Rather, it is "a total structure of actions brought to bear upon possible actions . . . a way of acting upon an acting subject or acting subjects *by virtue of their acting or being capable of action*" (1982, 789, emphasis added).

In the context of COVID-19, claims of a public health emergency did not, as Elaine Scarry (2011, 7) might contend, induce people to "stop thinking" or to "surrender [their] powers of resistance and forms of political responsibility." Rather, these forms of responsibility were redefined and amplified by digital platforms. Technologies like instant messaging empowered people with a sense of agency by offering the semblance of active choice, whether to download the app, check a website, or keep notifications enabled. This perceived agency to respond to crisis suggests a more collaborative and interdependent relationship between citizens and the state than is typically recognized (Trnka 2020). Indeed, such participatory approaches often foster sentiments of moral obligation and mutual support, challenging the conventional view that emergencies are solely opportunities for the top-down exercise of power.

Public responses to COVID-19 restrictions in Hanoi were dynamic, fluctuating along a spectrum from compliance to noncompliance, with many people navigating a middle ground. Individuals, myself included, might adhere strictly to measures one day only to bend the rules the next. This fluidity became increasingly pronounced as the pandemic progressed and perceptions of risk evolved. While highlighting the prevalence of consent, agency, and solidarity within collective action, it is equally important to acknowledge the role of dissent. As I explore in acts 2 and 3, dissent manifested through both subtle and overt acts of refusal, reappropriation, or disengagement, which ebbed and flowed across pandemic time. This nuanced understanding of people's actions does not ignore the emergence of new forms of "syndromic surveillance" carried out in the name of biosecurity and economic stability (Fearnley 2008). Rather, it illustrates the shifting boundaries and dynamics of care between state and society, accelerated by digital media technologies that compressed time and space in crisis management. Contact tracing apps like NCOVI and Bluezone, both domestically developed, were initially downloaded by millions, demonstrating a remarkable willingness to disclose sensitive personal data, including medical histories and location tracking, with government agencies. This involvement likely stemmed from the "promise of immediacy"—the allure of rapid access to "epidemic intelligence" that could enable early detection and swift responses to potential outbreaks (Caduff 2014, 34), aligning individual risk reduction with public health monitoring systems.

Syndromic surveillance in a digitally mediated pandemic operated both vertically and horizontally, tethered to expectations of reciprocity. This system facilitated "real-time surveillance" of self and others in exchange for "a sense of participation as well as control" (Engelmann 2020). Through live virus trackers and synchronized data fragments, users simultaneously shared bytes of personal information while staying informed of breaking developments, transforming epidemic events into networks of collectively mined, aggregated, and circulated knowledge about COVID-19. During the first months of the pandemic, Vietnam's Ministry of Health harnessed this digital ecosystem through targeted mass messaging, sending an estimated six billion notifications via SMS and popular apps like Zalo (Huynh 2020).[2] As smartphones continuously pinged and buzzed with updates, these messages established authoritative knowledge about viral threats while projecting transparency through direct and immediate contact with attentive citizens.

Text alerts and messages from authorities, as a technology of "real time biopolitics" to mitigate threats to collective life (Lakoff 2017, 99), are by no means a new mode of emergency preparedness and communication. During the pandemic, governments all over the world utilized instant messaging about COVID-19 to disseminate vital information, curb virus transmission, and implement practices of e-care, like in Vietnam. Despite being a routine public health strategy for reaching broad populations efficiently—the CDC itself provides guidelines for SMS best practices—Vietnam's use of these technologies during the pandemic was framed by Western media as digital authoritarianism. A commentary in an American news publication illustrates this interpretation, opening its critique of Vietnam's "repressive" disease prevention policy with an Orwellian scene: "When [a] Hanoi-based economic consultant . . . returned home after a trip abroad in late March, he was immediately texted by the local police asking after his health. Vietnam is a state that not only knows where you live but also knows when you go away—and your mobile phone number" (Hayton and Tro 2020). This framing reveals how neighborhood health protection measures were recast as invasive surveillance.[3]

The economist, a permanent resident of Hanoi, later challenged this dystopian portrayal, objecting to the article's misleading use of his tweets to advance its narrative. Instead, he acknowledged what he saw as an "impressive communications effort" by the government. Far from experiencing an abstract, omnipresent authoritarian state, he described a form of intimate governance that cultivated human connections between residents and municipal authorities. His arrival messages—two general pandemic guidance texts (similar to those I had received) and one personalized SMS from a local police officer inquiring about his family's well-being—were presented as examples of a more approachable style of public health management, one that humanized disease surveillance through the exchange of selfies, memes (including a pulsating heart), and emoticons.[4] By using friendly and relatable content, these messages aimed to convey care and attentiveness through a more empathetic approach to monitoring COVID-19. The pandemic thus opened up new channels for governance through affect and the scaled distribution of sentiment (Stoler 2004), a process enabled by the expansion of interactive digital platforms.

As with the economist, my familiarity with the state grew through instant messaging, particularly after being placed under quarantine, when I began to anticipate a barrage of daily pings, chimes, and buzzes on my device. Mindful of Scarry's (2011) warning that crises fueled by fear and uncertainty can suspend critical thinking about power, I found myself, as I often do in Vietnam, recalibrating my privacy threshold. I felt fortunate to be granted entry into the country when perceived risks from outside were deemed high. The sound of those initial notifications upon my arrival in Hanoi provided an unexpectedly reassuring comfort, signaling the presence of a functioning state that took preparedness, planning, and response to crisis seriously—a stark contrast to the place I had hurriedly left, where disaster governance had been largely dysfunctional. The upbeat tone of the messages did more than automate disease surveillance; they embodied an ethics of collective care central to public health security. As Tine M. Gammeltoft (2022) observed in other medical contexts in Vietnam, and as my research confirmed, care and control during COVID-19 were inextricably bound, working dynamically in tandem to manage risk and prevent the spread of disease.

ON BEING PREPARED: SECURITIZATION AS CARE

If emergencies serve as moments to implement new modalities of power and care to reaffirm legitimacy, then the COVID-19 pandemic created an opportunity for the expansion of digital governance to mediate relations between citizens and the state through technology. Anthropologists have long recognized the historical connections between disease control technologies and the uneven distribution of vulnerability during epidemics (e.g., Lynteris and Poleykett 2018). In Vietnam, national digital transformation (*chuyển đổi số quốc gia*), in accordance with the Fourth Industrial Revolution (*Cách mạng công nghiệp lần thứ tư*), or Industry 4.0, sought not only to optimize the efficacy and best practices of state agencies in the quest to develop a "digital society" (*xã hội số*). Decision 749/QD-TTg 2020, pursuant to Resolution 52-NQ/TW 2019, put into action a series of digital reforms to improve perception of government, while fine tuning the technological operations of everyday rule.[5] E-governance collapsed time and space through the speed and scale with which authorities could reach people—and people authorities—any place or time of the day. It also facilitated the transition from proactive disaster preparedness to a coordinated, rapid response as the potential threat of COVID-19 escalated to a full-blown public health crisis.

Even before the first cases of the novel coronavirus were reported in Wuhan, China, the Vietnamese government had meticulously planned for the next zoonotic epidemic. Swift and decisive action had successfully contained previous infectious disease outbreaks originating from interspecies transmissions, like severe acute respiratory syndrome (SARS) in 2003 and subsequent avian influenza (Porter 2019). Expectations of a likely disastrous event—a matter of *when* not *if*—meant that prevention, rather than relief, was the primary focus of national preparedness strategies to reduce impact on public health and the economy. Scholars have pointed to the logic of anticipation that drives preemptive action and risk-readiness rationalities. Anticipation of future emergencies serves to justify present securitization of society (Choi 2015, 290) while generating knowledge about existing vulnerabilities (Lakoff 2017, 19). Anticipatory interventions in Vietnam are shaped not only by the certainty of an uncertain future, but also by institutional memories of past catastrophes.

Analyses of disaster preparedness demand attention to political economy and the relationship between modes of production and emergency responsiveness to understand which people are valued in society and who are deemed disposable, existing solely to create surplus value (Wisner 1978, 81). Unlike "disaster capitalism," which unevenly and *reactively* outsources relief to private entities—resulting in catastrophic outcomes for poor and racially marginalized communities during Hurricane Katrina in 2005 (Adams 2013) and for the elderly during the 1995 Chicago heat wave (Klinenberg 2002)—"disaster socialism" emphasizes a strong public sector with robust response capabilities that prioritize collective life over individual autonomy. This approach continues to be effective in countries with substantial public investments operating under socialist frameworks (including socialist-oriented market economies), where government leads planning and infrastructure development. Cuba illustrates this well; its renowned hurricane preparedness system has maintained low mortality rates despite resource constraints (Sims and Vogelmann 2002), while its "people over profit" pandemic response demonstrated the human benefits of a "holistic and integrated" strategy, particularly in terms of lives saved (Hosek 2024).

Similarly, Vietnam's whole-of-society approach to disaster governance continues to place public health before economic growth, reflected in prevention and mitigation strategies that mobilize people and resources on a massive scale to minimize risk.[6] Historical memory, particularly of human-made disasters like warfare, has helped sustain commitments to social solidarity and volunteerism, qualities that strengthen efforts in strategic planning, coordination, compliance, and outreach. From the evacuation of millions of civilians during past aerial bombardments to current typhoon responses, often with military assistance, Vietnam's comprehensive strategy vividly illustrates how disasters become "securitized"— transforming concerns about public safety and care into matters of national security (Older 2016).

Preparedness for COVID-19 similarly emerged at the intersection between health and security concerns (Collier and Lakoff 2008, 14). Securitization was most evident at border checkpoints in Vietnam, where security forces, including customs officials, worked in tandem with public health workers to control the spread of communicable disease. Biodefense as active threat monitoring began at national borders, which slowly shut down around the world. During the pandemic, points of entry were perceived to be especially vulnerable "weak spots" for global health threats and microbial penetration through contaminated foreign bodies. To prevent the virus from entering at international gateways like Hanoi's Nội Bài Airport, authorities implemented universal testing and systematic data collection for passenger screening, enabling early detection and risk assessment to curtail community transmission. Through enhanced digital health governance, biorisk management at borders shifted the focus of control from contagion itself as the object of medical intervention to the potential carriers, such as people and objects, believed to harbor and spread the disease (Biehl 2016, 133–34).

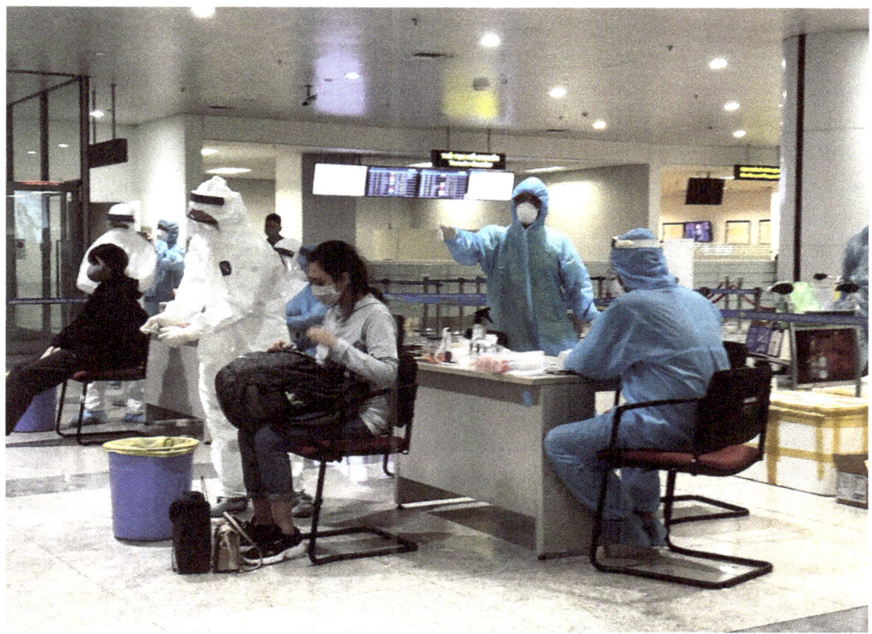

FIGURE 5. Securing borders: immigration hall repurposed as COVID-19 testing site for international arrivals, Nội Bài International Airport, March 2020. Photo by the author.

Digitalization strengthened national security capabilities through the e-governance of public health and care work from afar. Even before my flight to Vietnam, aviation policy mandated self-monitoring as a condition of travel and entry through an online medical declaration, or *tờ khai y tế* (figure 4).[7] Up to that point, my experience with Vietnam's early delivery of digital government services had been largely positive. The loose-leaf visas I obtained twice through online applications in 2019 and 2020 demonstrated the speed and convenience of securing entry authorization without having to surrender my passport. Predeparture submission of my health status and personal data—which the digital form assured me would be handled with "complete confidentiality" (*hoàn toàn bảo mật*) in accordance with communicable disease control guidelines—demonstrated to authorities my willingness to comply with pandemic regulations, including biosurveillance, to gain admission into the country. My self-reported absence of symptoms over the previous fourteen days indicated a low-risk profile, suggesting I was unlikely to pose an immediate danger to human life and public health.

Stepping into Hanoi on March 17, I encountered a transformed arrival space; the hum of medical equipment had replaced the usual bustle of travelers, and sanitized halls and corridors now served as makeshift medical facilities (figure 5).

This sensory-spatial reconfiguration embodied how public health had become a matter of urgent national security. Through the sights, sounds, and smells of this medicalized gateway, I experienced emergency preparedness and biosecurity as materialized through specific techniques of e-governance, while also recognizing their limitations. After disembarking from the plane at Nội Bài in a calm and choreographed fashion, one row at a time, I walked two strides ahead of the rest of the dozen passengers, maintaining my distance, until we reached the medical quarantine desk, the first control point. An airport employee dressed in a blue hazmat suit screened my body temperature as I held up the declaration QR code I had received digitally—TK-00011972—for her to see. The thermometer's shrill beep, an auditory confirmation that I would soon associate with everyday pandemic procedures, signaled my clearance to proceed.

As the signal from the infrared non-contact device faded, I was ushered to the next station. There, a health official scanned my QR code, sounding another electronic indication. Around me, other passengers, mostly repatriating Vietnamese, hurriedly filled out their paperwork. There appeared to be a lack of familiarity with e-government initiatives, I thought, upon realizing that I was the only person who had used the digital platform. The health official checked my online information, asking me again about any perceived symptoms, before printing a health verification certificate that bore my name, place of departure (California), and the QR code now linked to my well-being. With a decisive thud, the officer stamped "Đã Kiểm Tra" (controlled) in bold red letters. Handing me the document, which included a COVID-19 hotline, they gestured toward the next control point: the normally crowded immigration hall, now vacant and converted into a sprawling testing site and makeshift laboratory.

Technicians clad head to toe in protective gear escorted me to a seat, placing a paper consent form in my hands as they prepared the vials and swabs for nasal and oral sampling. One worker, their voice muffled by the mask, inquired about the number of cases in California, which had yet to reach one thousand. By the time I cleared immigration—typically a bottleneck but now expedited with a quick scan of my loose-leaf visa barcode—my fellow passengers were only just beginning their health screenings, a contrast to the efficiency I had just experienced.

As I descended the escalator to the baggage claim, a sharp buzzer pierced the air before the conveyor belt jolted to life. Quickly grabbing my luggage, I was waved through customs—clearly, no one was eager to handle potentially contaminated items—and joined my spouse in a taxi just outside. Only days later, Vietnam suspended issuing visas to foreigners. A jarring text alert that morning woke me to some news that would define the next year of my life: "EVA Airways informs: Flight BR398/06APR has been cancelled." There were no more flights out of Hanoi. On March 25, the government closed the borders to all international travel. Stranded, my return to California was indefinitely postponed.

THE COVID REVERSAL: RETHINKING THE SCARCITY SLOT

Innovative forms of digital communication—including those with sonic capabilities, such as viral videos and audio-based messaging, along with mass texting and online health declarations—served as media technologies served as media technologies for enhancing the political legitimacy of the party and state through persuasive claims of effective pandemic preparedness and outbreak management (Nguyen-Thu 2020, 145). Digitalization also afforded new forms of civic participation, and, consequently, increased state control over the flow of social media content, particularly when deemed "fake news" (*tin giả*). While the modes of electronic transmission to spread public health messages, secure public trust, and ensure compliance were not entirely new, the scale of their reach was unprecedented. A Facebook-commissioned survey in 2020 revealed that 92 percent of rural households in Vietnam had smartphones, with an equally high rate of internet penetration (90 percent), significantly higher than the commonly cited statistic of 70 percent of the population as active internet users (Nguyen and Ho 2020, 8). This is not to deny the existence of a digital divide—particularly between lowland and highland areas, as well as across gender and generation—but to recognize the extent to which digital communication has come to displace traditional mass media, such as television, more widely than previously thought.[8]

Vietnam's push to transform into a digital society while strengthening its information and communications technology (ICT) capacity as part of Industry 4.0 has created new opportunities for the growth of private enterprises, both small and large. E-commerce related to food in particular surged during the pandemic. Women, such as the Vietnamese baker in my neighborhood, had long used social networking sites, including Facebook, to sell homecooked meals and other goods. However, demand soared with the onset of COVID-19 as people sought greater convenience and protection from infection. Food delivery platforms, like GrabFood, became the main income generator for motorbike taxi drivers who lined up outside food stalls and eateries that had pivoted to takeaway services, waiting to transport meals rather than riders. Despite government initiatives to reduce disposable, single-use plastic waste, the ecological impact of this online ordering and delivery industry, which was vital for keeping family-run businesses afloat during lockdown, could be seen, smelled, and heard in garbage collection carts around the city.[9] This included the rusted neighborhood cart outside my apartment in Ba Đình District, where piles of plastic and Styrofoam packaging had

FIGURE 6. The new shortage economy, California supermarket, March 2020. Photo by the author.

been tossed alongside discarded materials bypassed by female waste pickers, who perform essential but unpaid urban care work (Corwin and Gidwani 2021). Each night, municipal garbage crews hauled off the accumulated waste, the crunching sound of their truck's compactor shattering the stillness of curfew.[10]

Unlike in the United States, Vietnam did not suffer from critical shortages of consumer goods or everyday medical supplies in the early stages of COVID-19. Before my departure from California, my early pandemic experience was marked by mounting scarcities and unpreparedness—empty store shelves, rising social anxiety, and waves of panic buying in anticipation of calamity. First was the lack of protective masks and sanitizers as people scaled up individual readiness, followed by the hoarding of tissue paper products. None of this alarmed me initially. I concocted my own lavender-scented disinfectant with rubbing alcohol I found in the cupboard, and my student, who was conducting research in Vietnamese nail salons in Orange County, secured some face masks, rubber gloves, and a half-used bottle of hand sanitizer for me. But then the shortages grew from hygiene supplies to nonperishable goods, like canned beans, and then to dry goods like flour, eventually expanding to perishables, including meat and dairy products. Empty shelves stood as visual signs of impending disaster (figure 6). Their flour-dusted surfaces hinted at a mad dash to stockpile, raising ethical concerns about consumer conduct. Merchants responded quickly by placing limits on the purchase of certain products to ensure rational consumption and fair market practices.[11] Pharmacies moved coveted items behind the counter to curb voracious buying, while supermarkets imposed purchase limits. Cashiers, newly tasked with regulating access to basic necessities, were stationed near bold-letter signs declaring:

DUE TO HIGH DEMAND LIMIT OF 3 ON COLD & FLU MEDS, HAND SOAP, VITAMINS, HAND SANITIZER, HOUSE HOLD CLEANING (INC. AIR CARE), WATER, BATH TISSUE, PAPER TOWELS AND FACIAL TISSUE.

Shortages introduced a radically different temporality in the United States as the rhythms of everyday life abruptly changed to that of time-space *decompression*, marked by experiences with delays and disconnections. As disruptions to supply chains and logistics curtailed commodity availability, Americans encountered a temporal condition imagined of dysfunctional *elsewheres*: that of waiting. Waiting, Cheryl Mattingly (2019) argues, is inherently anticipatory. One remains, lingers, dwells in a particular meantime, suspended in a state of deferred action—of desiring, but not being able, to move forward. During the pandemic, people in the United States waited for goods, from household essentials to luxury items, that took longer to arrive. They joined waitlists. They stood in line—in *queues*. The queue: that quintessential sign of alterity, of the specter of *socialist* temporality and command economies rather than of global capitalism. As Sarah Sharma has astutely observed, "meantime temporalities" and experiences with waiting are problems of time typically ascribed to *others* (2014, 52). The pervasive image of the socialist queue—and of docile bodies waiting in vain for unattainable, if not unspecified, commodities—came to define for the capitalist West what anthropologist Krisztina Fehérváry (2009) has called the "shortage paradigm" as the dominant experience of everyday socialism and socialist state bureaucracy.[12] And yet empty shelves during the pandemic served as a reminder of the scarcity that capitalism itself produces and depends on for its value (Giles 2021).

Vietnam has long occupied the "scarcity slot" (Logan 2020) in Western imagination. Common depictions of the war-ravaged country as static and unchanging until "saved" by market reforms after the collapse of communism paved the way for a "we-won-the-war-in-the-end" rationality in the United States (Schwenkel 2009a, 202). Of course, the history of food and material insecurity in Vietnam cannot be attributed solely to the failed policies of a centrally planned, "shortage economy" (Kornai 1980). The legacies of empire and colonial resource extraction, targeted bombing of infrastructure, and a punitive trade embargo lifted only in 1994 ensured that poverty and deprivation remained the unnatural order of things. Though food rationing and the subsidy system ended in 1986 with the launch of Đổi mới reforms, pharmaceutical shortages persisted into the early 1990s (Craig 2002, 125), as did issues with access to potable water and stable electricity in secondary cities (Schwenkel 2020, 248). Like other postcolonial nations, Vietnam's underdevelopment allowed affluent countries in the Global North to position themselves as saviors of a "backward" population plagued by lack and scarcity. Their promotion of neoliberal policies aiming to stimulate "growth" sought to overcome what they perceived as radical temporal alterity—that is, to transform Vietnam into a modern, consumer-driven society through the production and accumulation of things.

As the pandemic unfolded, Vietnam and the United States found themselves in the unique position of scarcity and abundance reversed. While the United States grappled firsthand with an unforeseen shortage economy (Belsie 2021), Vietnam, a low-cost sourcing country, continued to optimize productivity, resulting in its own steady supply, if not excess, of commodities such as surgical masks. Donations of millions of PPE to wealthy, pandemic-stricken countries—including to the United States—as part of its "coronavirus diplomacy" flipped the donor-recipient hierarchy that has traditionally defined Vietnam's relationship with the Global North. This symbolic inversion of the shortage paradigm cast doubt on the United States' claim to being the world leader in science and humanitarian assistance, as nations like Vietnam emerged as "pandemic winners." Anthropologists recognize that conceptions of scarcity and abundance are culturally and historically contingent, relational, and political (Scoones et al. 2019), that is, reflective of particular power interests. In line with this understanding, Cold War experiences of scarcity in Vietnam were relative rather than absolute, varying by time and place; consequently, the socialist East emerged as a consumer "paradise" for Vietnamese workers and students abroad (Schwenkel 2022a). Still, I had not anticipated the ready availability of essential commodities during the pandemic, particularly the surplus of hygiene products and protective equipment. Face masks, deemed vital for disease prevention, could be found across retailers and pharmacies, where they sold for mere pennies.

Under crisis conditions that heighten supply chain vulnerabilities, ethical consumption requires careful consideration of what to buy, where, and in what quantities, while remaining mindful of the needs of others. Fresh from California's acute market shortages, I worried that my spouse and I, as foreign guests in Vietnam, might strain local resources in the event of wider outbreaks. Could Vietnam also face scarcity due to disruptions to logistics operations caused by border closures? We were unsure, so we adjusted our purchasing habits accordingly until we gained a clearer sense of the developing situation. Instead of shopping at the crowded yet convenient wet market down the street, our usual spot where my neighbors bought fresh food daily, we traveled across town on our motorbike to a small gourmet shop with imported goods, assuming there would be less demand. We also frequented a well-stocked hypermarket that offered ample space for social distancing and attracted middle-class Vietnamese customers seeking a wider selection of food options.

As it turned out, my assumptions proved mostly unfounded. The pandemic quickly reshaped daily consumer practices as people assessed risk of infection with the need to procure food for their families. The Châu Long wet market in my neighborhood, for instance, was a largely unregulated space of commerce with a semi-open-air design that blurred the line between inside and outside, allowing customers on motorbikes to buy goods while driving past stalls. That very fluidity increased people's anxieties about entering less hygienically controlled spaces where masked female vendors handled fresh meats and produce with their bare hands. Reports that the novel coronavirus had originated in a wet marketplace in Wuhan further contributed to the atmosphere of unease, even among the traders themselves. When I first visited the wet market after my return to Hanoi, I found a markedly subdued and altered sensory atmosphere. Several stalls were closed, including my usual vegetable stop (due to "cô vy," or COVID, the neighboring trader informed me). Reduced visitor traffic also meant the usual lively sounds of hawking, chopping, and chattering across rows of vendors were greatly diminished.

Yet the melodic cries of itinerant traders outside the market, a sonic fixture in the city, offered a counterpoint to the market's muted energy. As they peddled down my street on bicycles equipped with battery-powered speakers, calling out, "Who wants hot, crisp bread?" (Ai bánh mì nóng giòn đây?), my neighbors rushed outside to buy baguettes, keen to avoid lines at local bakeries.[13]

Supermarkets, on the other hand, were controlled environments. Despite the higher prices, they were more crowded but oddly quiet as people shopped quickly without lingering to examine or discuss merchandise with fellow shoppers. To gain entry, customers underwent a standard screening procedure performed by security-guards-turned-essential-workers using contactless thermal technologies. First was the digital temperature check, which emitted a long mechanical "beep" when complete, followed by the short chirp of an automatic dispenser that sprayed disinfectant onto hands. Bright lights, high ceilings, and polished tile flooring conveyed an aura of cleanliness and safety absent in the wet market with its slippery, concrete surfaces, low makeshift roof, and dimly lit stalls in the evening. The spacious rows of packaged foods and household goods, stocked high on shelves, conveyed an impression of material plenty and logistical normalcy that somehow calmed me. That calm—really, a sense of relief—soon gave way to feelings of guilt, however, as I watched from afar the pandemic's devastating consequences unfold in the United States. The displays of abundance in shops transformed everyday commodities—like hand sanitizers of all sizes (figure 7)—into "melancholic objects" (Navaro-Yashin 2009) as I contemplated the shortages that accompanied the sharp rise in the number of infections and deaths back home. I took a photo of the bounty and sent it to my father. He texted back almost immediately with a hint of despair: "Can you send us some masks?"[14]

This seemingly simple request proved unexpectedly challenging, and not because border closures had disrupted delivery systems or the global movement of goods. Readied with a securely sealed yet lightweight envelope, I ventured out to the courier company close to my apartment. The employee eyed my care package

FIGURE 7. Stocked hand sanitizer shelves on display in a Hanoi supermarket, March 2020. Photo by the author.

with suspicion. "What's inside?" she asked promptly while scanning shipments piled on the floor. "A few masks for my father; he lives in the United States," I answered truthfully, searching for a customs declaration form. Surely she was aware of the shortage of PPE abroad, widely reported in the local press, which reversed perceptions of US plenitude and prompted the Vietnamese government to consider sending medical aid to its former adversary, enhancing its position on the world stage. She stopped and looked up, shaking her head. "Not allowed" (*không cho phép*), she said curtly, and upon noticing my puzzled face, explained that the government had banned the export of masks (and rice) to protect domestic supply chains.[15] A regulatory framework designed to control for scarcity ensured that market supplies and distribution networks remained intact, safeguarding food security—especially for those in centralized quarantine.[16]

SOLIDARITY ECONOMY

It was a late Sunday afternoon at the end of March when my phone chimed with a digital fundraising text from the charity platform 1400.vn:

To support the prevention and control of the COVID-19 pandemic, please text:
CV n, send to 1407 (In which, n is the number of times to support 20,000 VND).
Each message you contribute 20,000 VND multiplied n times (where n is limited
from 1 to 100). Telephone support: 19001530, extension 6 (1,000 VND/minute).
Website: 1400.vn. Thank you.[17]

With few reserves in my account, I texted back to 1407: "CV2," and received an immediate response: "Thank you for your support of 40,000 VND [US $1.80] to fight against COVID19."

The National Humanitarian Portal 1400 (Cổng thông tin điện tử nhân đạo quốc gia 1400), managed by the Ministry of Information and Communications, deploys new media technologies to extend what might be considered an infrastructure of solidarity to cell phone users nationwide. Its motto—"Let love stretch far" (Cho yêu thương vươn xa)—mobilizes affect and memory to foster networks of care and a sense of national belonging through mobile donations, while invoking Vietnam's tradition of "remembering the source of the water that one drinks."[18] Text message–based charitable giving represents an innovative form of digital philanthropy, grounded in principles of reciprocity and redistribution that reflect deep-seated moral and social obligations. Text numbers are pegged to specific donation campaigns, many of which are continuous, ebbing and flowing across seasons, such as 1409 for Agent Orange victims, 1403 for flood losses, and 1404 for Trường Sơn (Hồ Chí Minh Trail) veterans. The "people's" (toàn dân) campaign to prevent the spread of COVID-19, which ran for three months from March 19 until June 18, 2020, raised 152.3 billion VND (approximately US$6.7 million) from 2.6 million subscribers, garnering more donations than any other fund drive that year by over a hundredfold.[19]

Disasters can act as powerful catalysts for social change, providing vital opportunities for meaningful expressions of care and altruism that strengthen social cohesion during crises (Solnit 2009). Recent attention to "pandemic solidarities" in particular has highlighted the creative efforts of mutual aid networks that arose around the globe, independent of institutionalized power structures (Sitrin and Sembrar 2020). Narratives of volunteerism and selfless support for the greater public good were also prevalent in Vietnam. Among the most well-known examples, widely covered in the press, were the innovative "rice ATMs" (ATM gạo) and "mask ATMs" (ATM khẩu trang), which distributed free essential goods to vulnerable residents facing hardship.

While these popular pro-bono initiatives were presented as independent charity work spearheaded by a young, aspiring CEO in Hồ Chí Minh City (whose model spread to other provinces), they operated less outside government institutions than in coordination with them. Similarly, at the grassroots level, as neighborhoods organized within local ward structures, these endeavors, too, reinforced prevailing ideologies about how to be a responsible, enterprising, and morally virtuous citizen. During the pandemic, public displays of social responsibility—shown through acts of care and reciprocity—fueled a vibrant solidarity economy that filled gaps in state support. This system promoted an image of national unity by emphasizing collective duty and sacrifice, with each person contributing, whether in small or significant ways, to overcoming (*vượt qua*) a public health emergency through the redistribution of wealth.[20]

Crises not only generate solidarities to ensure collective well-being but also require the scaled allocation of responsibilities across various domains of social and political life, involving both state and non-state actors. Susanna Trnka and Catherine Trundle note that responsibilized subjects "exist within a matrix of dependencies, reciprocities, and obligations" that govern relations, including those between citizens and the state; thus, their actions cannot be solely understood through neoliberal logics alone (2014, 150; see also Watanabe 2021). This dynamic of distributed rather than individualized responsibilities was evident in the discourse on remote learning during the pandemic, where shared burdens were emphasized. While parents primarily managed the logistics of distance education for their children—such as acquiring necessary technology, which posed a challenge for low-income families—the issue was framed as a public concern rather than a private family matter. One late afternoon, my phone pinged, alerting me to a new text, this time from the Ministry of Education and Training (MOET):

> BOGD-DT [MOET]
> April 3, 2020, 16:23
> To implement the slogan "pause going to school, but don't stop learning," MOET has announced the streamlining of the 2nd semester curriculum and guidelines for teaching/learning via the internet and television. Teachers, families, and society as a whole should join hands to support, motivate, and help pupils study through the internet and television to ensure accessibility, quality, and efficacy.[21]

Overlapping forms of relational and social contractual obligations, extending to citizens of all ages (including pupils), shaped the solidarity economy, with the state at its helm. During the pandemic, the state seized the opportunity to reassert its role as the custodian of safety though collective biopolitics, even as it redistributed duties. This approach centered on administering collective life to sustain both health and society through coordinated interventions that tracked, contained, and regulated bodies, all guided by an ethos of the common good. As Latour (2020) observed, "Pandemics awaken in leaders and those in power a kind of self-evident sense of 'protection'—'we have to protect you' 'you have to protect us'—that recharges the authority of the state" and its relationship with the governed.

Consider quarantine, a classic biopolitical tool of spatial control (see Foucault 1977, 195–97) that also functions as a mechanism for governing time. The collective quarantine (*cách ly tập thể*) of entire neighborhoods, streets, and apartment complexes

FIGURE 8. Enacting state care: People's Army and volunteer defense forces (with caps) distribute food to quarantined residents on Trúc Bạch Street, March 2020. Photo by the author.

cordoned off with red tape was framed as benevolent governance necessary to safeguard the city. In my ward, where an entire block was placed under quarantine after the first superspreader event (see pandemic figure 1), state rituals of care mobilized the solidarity economy to regulate the rhythms of daily life, ensuring physical and social well-being. Lefebvre reminds us how the "rhythmed organisation of everyday time" is at once internal and external, both individual and social (2013, 75), and in the case of collective quarantine, rendered governable as a matter of biosecurity. At carefully scheduled intervals throughout the day, teams comprising self-defense forces, health care practitioners, municipal authorities, and military personnel coordinated a complex care operation for quarantined residents. At predetermined times, residents emerged from their homes to collect food rations, gift packages, and essential items, or to undergo medical checkups and COVID screening—events often observed by state media and occasionally the anthropologist (figure 8).

In the late afternoons, the desolate street came alive as families were allowed outside for their allotted fresh air and exercise, a routine designed to facilitate health monitoring from a distance.[22] The laughter of children riding bikes mingled with the metallic clangs and creaks of the exercise equipment being used by adults along the canal behind my building. These sounds drifted up to my office window, soon to become my quarantine outpost. From there, I would experience firsthand

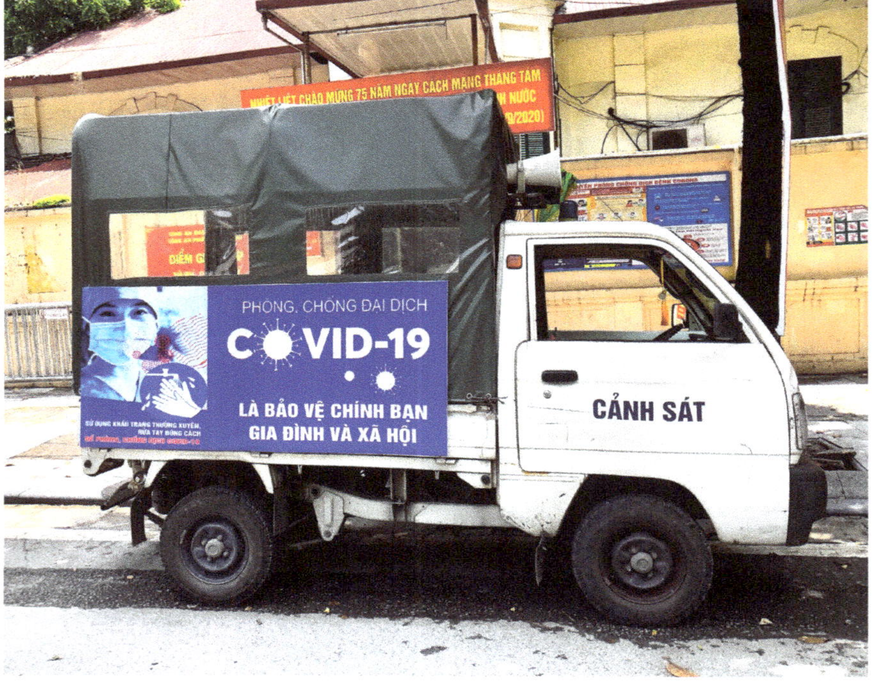

FIGURE 9. "Prevent COVID-19 by protecting yourself, your family, and society": roving police sound truck doubles as public health vehicle in Hanoi, August 2020. Photo by the author.

the orchestrated routine of pandemic governance as it monitored the rhythms of my own daily life.

QUARANTINE: CARE AND CONTROL

I spent my first days in Hanoi attuning myself to an unfamiliar soundscape punctuated by both sonic absences and augmentations. The suspension of in-person instruction had silenced the once boisterous, morning assembly of children and teachers in the courtyard of the Vietnam-Cuba primary school adjacent to the Châu Long market. Meanwhile, new sounds emerged, including fleets of roaming sound trucks equipped with public address systems, broadcasting COVID-19 health messages into the privacy of homes (figure 9). Everyday rituals of public safety likewise reshaped the acoustic landscape. The affirmative "beep" of digital thermometers wielded by security guards, or người bảo vệ, became a persistent auditory feature. These essential yet vulnerable workers evolved into powerful arbiters of public space through their frontline role in disease detection, controlling access to businesses and government offices while redrawing lines of authority (see pandemic figure 2).

By the time I arrived, my landlord had already retreated to her family's *quê hương* (home village). The exodus of city dwellers reflected a perception of the virus as an urban malady, echoing historical views of the dense capital city as a vector of disease.[23] Thanks to the expansion of municipal e-government, my landlord was able to manage my re-registration with the local police through digital documentation, sparing a return to what was then considered ground zero (Trúc Bạch Ward). This streamlined process proved remarkably efficient. Shortly after she submitted my visa electronically, I returned from an early morning walk around the lake to find the building's *người bảo vệ* waiting with a set of official papers. Each document was meticulously signed in blue ink and adorned with an official red stamp, informing me that I was quarantined—effective immediately. To my surprise (and some relief), the mandated two-week confinement was applied retroactively to my March 17 entry date, ending on March 31. This unexpected consideration shortened the duration of the time I would spend housebound under the medical oversight of an attentive state nurse.

From the administrative paperwork presented to me—including the Decree on Home Quarantine (containing my biodata, some of it incorrect), the Residential Quarantine Guidelines, the Quarantine Pledge (signed and stamped but not yet filled out), and the Disease Prevention pamphlet—I learned that pandemic governance was a tangible lesson in Vietnamese bureaucracy.[24] While some of these documents were available online as part of e-government procedures, their printed versions demonstrated how a "political economy of paper" (Hull 2012) continued to play a crucial role in Vietnam's health care infrastructure during times of crisis. As bureaucratic artifacts with rich ethnographic value, the documents rendered the abstract state and its pandemic preparedness more legible, materializing how governance operates in a public health emergency. For example, they outlined lines of communication and institutional coordination, specifying assigned duties across levels of government—national (*quốc gia*), municipal (*thành phố*), district (*quận*), and ward (*phường*)—and across multiple agencies, from health departments and security forces to party councils and culture-information units. This hierarchical structure connected high-ranking officials like ministry heads (*bộ trưởng*) and People's Committee chairpersons (*chủ tịch UBND*) with community-elected neighborhood leaders (*tổ trưởng dân phố*). These street-level bureaucrats, operating at the lowest tier of governance, proved most impactful, serving as bridges between policy and public safety by monitoring the health and welfare of households under their jurisdiction.[25]

Paperwork mediated my relationship with the state, making it more intimate and familiar in a manner that felt nonthreatening. The documents I received were an invitation to compliance without the need for coercion; they stipulated in precise bureaucratic language the code of conduct expected of me while under the direct care and supervision (*giám sát*) of district and ward health care workers. This included ethical commitments (not vacating the dwelling without permission or concealing any symptoms), adherence to hygiene measures (ventilating the home, disinfecting surfaces, separating waste, and washing hands), and observance of distancing requirements (eating and sleeping in a separate room, or, if the residence was too small, retiring to a separate bed while masked). These guidelines aligned with the five measures mass-texted frequently by the Ministry of Health to promote hygienic sensibilities as central to civic duty: stay indoors, wear a mask and keep a safe distance if going out, wash hands, sanitize surfaces, and update medical declarations in order to ensure healthful living (*sinh hoạt lành mạnh*) and wellness tracking (*theo dõi sức khỏe*) by authorities.[26]

Quarantine paperwork personalized state care by placing my health (and personal data). Into the hands of specific medical practitioners. These individuals, referred to as "comrades" (*đ/c*, short for *đồng chí*), were listed along with their personal cell phone numbers, in the documents. This included Comrade Hồng, the head of the medical station for Trúc Bạch Ward, and Comrade Hương, the nurse responsible for performing my twice-daily digital health checks. While Hương's affective labor served clear biopolitical goals—managing contagion risks to safeguard collective life—our ten-day virtual relationship, centered on monitoring my health, revealed how state modalities of (e-)care penetrated and governed the intimate spaces of the home.

Our hyper-mediated exchange unfolded through the digitized sounds of instant messaging (IM), illustrating the role of media technologies in shaping a sense of place (Pink 2015, 121). The platform enabled contactless, near-real-time surveillance—no touch, no sight, no risk of exposure to disease. Hương established our communicative pattern with her first text that afternoon, announced with an abrupt ping that had me scrambling for my phone: "Có vấn đề gì về sức khỏe chị cứ gọi cho e nhé" (If you have any health issues, just call me, ok?). Her use of relational honorifics—"chị" (commonly translated as older "sister" or senior female person) and "e," the IM short form for "em" (signifying a younger sibling or junior person without reference to gender)—established a respectful yet informal tone. The exclamatory "nhé" at the end of the text conveyed a friendly, almost playful mode of digital interaction, quite different from the reserved communication I had come to expect from health practitioners. Rather than issuing instructions, this affective resonance established a tone of encouragement, presenting medical supervision more as a practice of relational care than an exercise of authority.

Hương's message left me with ambivalent feelings of security. The accessibility of a health care provider during a potential medical emergency was reassuring at a time when the world seemed to be unraveling. Her technological "presence" thus ensured my acceptance of digital surveillance. At the same time, the sound of her text heightened my anxiety about time and state oversight. "Got it, thank you em," I typed back with a hint of uncertainty, wondering if my response had been prompt enough to indicate my cooperation. Later that evening, I missed a text informing me that Hương and her team would visit in the morning to collect a respiratory sample, my second polymerase chain reaction (PCR) test within a week.[27] I confirmed and turned up the phone's volume. Without delay, another text alert sounded. "E cảm ơn" (I thank you), Hương wrote back, signaling her round-the-clock work to monitor viral spread.

The next morning, a loud humming sound announced the arrival of the medical team, followed by an overwhelming smell of chlorine. I opened the door to find a group of three individuals dressed in blue full-body protective suits spraying down the hallway with disinfectant, an act that, in Mary Douglas's terms, sought to unleash purity against danger.[28] Sterilization of surfaces, human and nonhuman, was common practice in Vietnam at the time, shaped by scientific assumptions about the risk of coronavirus transmission through surface contamination. "Disinfection and treatment of the environment" (*việc khử trùng, xử lý môi trường*) was encouraged by the Ministry of Health, as stated in its March 25, 2020, communiqué 1560/BYT-MT, following the first outbreaks in Hanoi, specifically in Trúc Bạch Ward on March 6 and Bạch Mai hospital on March 18.[29] Mobile units from the army's Chemical Corps (Bộ đội Hóa học) and the Defense Ministry's Chemical Battalion (Binh chủng Hóa học) deployed across the city in a mass decontamination operation in response to these outbreaks, their sprayers hissing forcefully as they saturated the streets with disinfectant.

Alongside such sonic amplifications, the disinfection of the city shaped a distinctive pandemic sensorium through pervasive chemical smells that registered at multiple scales. At the micro level, objects underwent sanitization to prevent *surface* transmission of the virus: jugs of drinking water were sprayed with bleach before delivery; hands were sanitized, sometimes involuntarily, upon entering buildings. At the macro level, the focus shifted to atmospheres, with outdoor spaces like streets and parks treated with chloramines to curb *airborne* viral transmission. The Ministry of Health eventually discontinued this widespread sterilizing practice, citing concerns about its efficacy and its potential harm to human health and urban ecosystems.[30]

During the state-mandated health visit to my home, the disinfection technician, outfitted with a backpack sprayer, thoroughly soaked the stairwell, causing me to worry about the neighbors. I imagined they had watched with concern as the blue team arrived, a sure sign of another outbreak. Meanwhile, two medical practitioners entered the flat, one carrying what looked like a small cooler. Curiously, neither identified as Hương, who, I was informed, was needed elsewhere. One practitioner prepared the paperwork for me to sign while the other administered the test. I winced in discomfort as they collected nasal and throat swabs before depositing them into the red chest (figure 10). The process was swift; they were polite but clearly eager to leave a site of possible contagion. They did not

FIGURE 10. Health screening and assessment by Trúc Bạch Ward medical team at start of home quarantine, March 2020. Photo by the author.

inspect the apartment for separate living quarters, nor did they seem concerned that my spouse was present and filming the procedure. I signed the quarantine and hygiene commitment both as the individual being quarantined (*người được cách ly*) and as the household representative (*đại diện hộ gia đình*). This pledge, along with its legal consequences for noncompliance, applied to the household unit rather than the individual, highlighting the collective nature of community-based medicine in Vietnam, where families bear the primary responsibility for health care (Gammeltoft 2014).[31]

Before departing, the practitioners provided me with a thermometer and instructions: I was to text my temperature to Hương twice daily, once in the morning and again in the late afternoon. They emphasized I should contact the ward health station immediately if I developed any symptoms—a fever (above 37.5 degrees Celsius or 99.5 degrees Fahrenheit), cough, sore throat, runny nose,

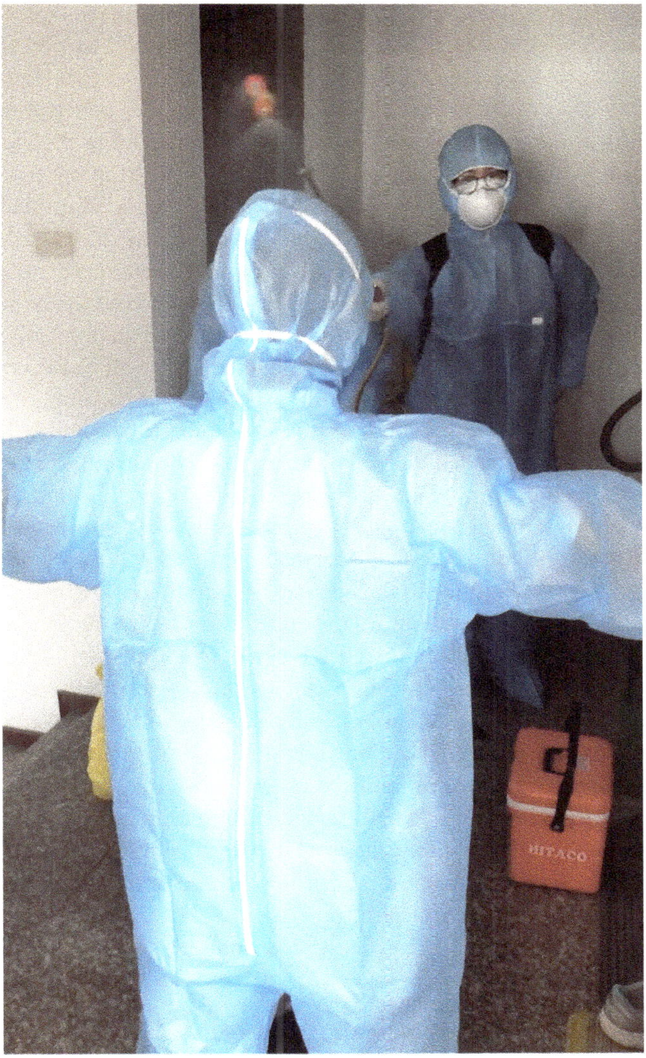

VIDEO 1. Trúc Bạch Ward medical team undergoes decontamination, adhering to safety protocols after completing home visit duties, 2020. Video by the author.

To watch this video, scan the QR code with your mobile device or visit
DOI: https://DOI.org/10.1525/luminos.249.video1

or breathing difficulties. After agreeing to these procedures, I escorted them to the door. There, they were met by the disinfectant technician, who started the engine and extended the vibrating wand. Each practitioner turned slowly, arms outstretched, as they were sprayed from head to toe (video 1). The loud mechanical humming ceased, but a sharp chemical odor lingered, penetrating my mask and irritating my eyes in the poorly ventilated space. After the elevator door closed, I went inside and waited a few minutes before texting my temperature to Hương: 36.3 degrees (97.3 degrees Fahrenheit). She responded immediately with "Ok, em cảm ơn nhiều" (Ok, I thank you very much), using the English loanword.

The pandemic introduced unprecedented forms of state governance and care that required refined sensory awareness, particularly of sound. As both a medium for managing bodies and information and a tool for navigating isolation, sound took on heightened significance. Quarantine, a condition of enforced separation from usual sensory stimuli to control risk, exemplified this complex relationship between isolation and sensory modification. Confined to my home, I found that perceiving the city demanded deeper attunement to the surrounding sonosphere, as active listening transformed into a vital practice for being in and making sense of the outside world (Janus 2011, 184–85).[32] This reorientation resonates with Jean-Luc Nancy's (2007, 4) critique of Western ocularcentrism: What does it mean to be fully immersed in listening, attuned to the nuances of sound? To be "formed by listening or in listening" as one "strains" toward meaning, toward sense-making or *hearing*?[33]

My quarantine sensibility rendered me "*ganz Ohr*" (all ears)—attentively listening both *to* and *for* sounds, while isolated from society. Nancy discusses the weaponization of listening and its early association with military surveillance, where one listened to secrets without being perceived (2007, 6). However, my quarantine listening was more anticipatory than reactive, given the scarcity of sounds to overhear from my seventh floor flat. Listening in anticipation of sonic events structured my daily life, synchronized to the rhythm of digital notifications and communication from nurse Hương. Temperature reporting demanded time discipline and consistent health tracking. Twice daily, at 8 a.m. and 4 p.m., I obediently sent an instant message with an outbound "swoosh" and then waited for the incoming alert:

March 25, 2020, 08:02

Hello em. This morning my temperature is 36.3
[Xin chào em. Sáng này NĐ chị là 36,3] *swoosh*

ping Good morning chị [Chào buổi sáng chị]
Thank you much [Cảm ơn iu]

March 25, 2020, 15:51

Hi em. My temperature is still 36.3
[Chào em. NĐ chị vẫn là 36,3] *swoosh*

ping Greetings chị [Em chào chị]
Yes, thank you chị (polite form) [Vâng chị ạ]

March 26, 2020, 08:04

Hi em, my temperature is 36,5
[Chào em, NĐ chị là 36.5] *swoosh*

ping Yes, thanks so much [Yes, thanks so much]

Anticipatory listening subsequently assumed dual roles during my confinement: serving as evidence of my compliance with public health regulations while also holding considerable personal significance. The chime of incoming text alerts, which I eagerly awaited, became particularly meaningful, as each sound marked a moment of social and emotional connectedness.

IM notifications punctuated my days with sound while introducing an unexpected informality to state-managed health care. Our writing style, while professional—occasionally using polite forms of address like *vâng ạ*—was surprisingly affable, defying the formal conventions that one might expect of official correspondence between practitioner and patient, state and subject. The vernacular medium of texting altered the tenor of our exchanges from the distant, impersonal style typical of state bureaucracy into something more colloquial and intimate. Hương often used short forms similar to those of my Vietnamese friends, casually incorporating English words or foreign salutations uncommon in Vietnam, like "good morning." Especially when texting outside our regular report times, she employed more familiar terms of address:

March 23, 2020, 10:27

ping Hey chị. I have a small favor to ask, can you text me,
chị, the flight number you traveled on. thank you
[Chị ơi. Em nhờ chị chút, chị nhắn cho em số hiệu máy bay
mình đi với ạ. e cảm ơn]

BR 397 từ Taipei đến HN em [BR 397 from Taipei to
Hanoi em] *swoosh*

The anticipated sounds of message alerts—outgoing swooshes and incoming pings—shaped how I lived and monitored my health under the care and control of an unseeing and unseen state. (After all, my only direct interaction with authorities was through individuals dressed in blue protective suits, devoid of any identifying markers.) In the absence of direct observation, I counted the days to my quarantine release with equal anticipation. On the afternoon of March 31, I sent my final temperature reading to Hương, slightly earlier than usual. She promptly responded, as always, but this time with a surprising note of praise: "Congratulations on successfully fulfilling the assigned task," she wrote, followed by a playful "hi hi" (hee hee) that gently mocked her own use of bureaucratic language. After I thanked her for her support, she replied with a trio of laughing face emojis. The cartoonish eyes and smiles beaming out at me, conveying satisfaction with my completed health tracking, infused our exchange with a layer of warmth. Through new media infrastructures, digital attentiveness evolved into an expression of e-care, fostering unexpectedly affective connections between citizens and the state in crisis.[34]

My quarantine obligation ended at 11:59 p.m. on March 31, 2020. Just one minute later, at midnight April 1, 2020, the country entered preemptive lockdown. As I detail in the next act, this transition prompted a shift in my sonic sensibility. As techniques of individualized *state care* evolved into an emphasis on collective *self-care*, my listening practice changed from anticipatory listening *for* to active listening *with*.

Domestic Workers

The Housekeeper (Người Giúp Việc)

The pandemic exacerbated gender disparities in uncompensated caregiving responsibilities within heteronormative households worldwide, as feminist scholars have noted. These disparities varied across race, class, and nationality, with working-class women and women of color facing particularly acute challenges as they balanced inequitable paid work with unpaid domestic duties while caring for children and adult dependents in extended families. These women were also more severely impacted by job losses than men (Bolis et al. 2020).

Despite widespread documentation of these gendered inequalities in caregiving duties, pandemic-era research predominantly captured care workloads in urban middle-class families, largely due to fieldwork constraints during stay-at-home orders. In Asia, attention centered on the strategies educated urban women used to manage without household employees during early lockdowns amid changes in the division of household labor (Uddin 2021), leaving a significant gap in knowledge about informal wage earners in the care economy. Initial findings revealed that women in precarious employment outside formalized systems experienced heightened vulnerability throughout this period. Live-in domestic workers were especially at risk, as seen with isolated migrant workers in Singapore, who were more susceptible to abuse, surveillance, and exploitation while confined to their employers' homes, with no opportunities to leave or socialize with co-ethnic peers (Antona 2020). They also faced greater risks of COVID-19 exposure from employers, as occurred with the household help who worked for Patient 17 (see pandemic figure 1).

These patterns of amplified precarity among workers in the informal care sector manifested with particular intensity in Vietnam. Like other countries that implemented strict coronavirus measures, Vietnam banned travel during its nationwide lockdown beginning April 1, 2020. Workers deemed nonessential to production and public services were required to stay at home. This included household help, or *người giúp việc* (person who assists with work), also known by the more gendered and condescending term *ô-sin*, borrowed from the title of a popular Japanese drama about a young girl working as a servant for affluent families. While many domestic care workers in Hanoi have been women who migrated from neighboring provinces to serve as live-in helpers (Nguyen 2010), others were "live-outs" who commuted daily to their employers' households. As urban families, rich and poor alike, hastily departed the city to return to their rural hometowns before the travel ban took effect, domestic workers confronted a dire loss of employment and income. Pandemic measures had already barred housekeepers from entering quarantined homes, affecting live-out workers like our weekly help. While we continued to pay wages throughout the lockdown, other *người giúp việc* were less fortunate. Moreover, those who managed to retain their live-in positions often endured exploitative conditions through increased workloads owing to mobility restrictions.

This situation played out in my apartment building in central Hanoi. My downstairs landlord and her family evacuated to the countryside in their black SUV, as rural areas were considered safer havens from what was then viewed primarily as an urban disease. They left behind Hùng, the live-in security guard or *người bảo vệ* (see pandemic figure 2), to assume a broader caretaker role when their housekeeper was prohibited from traveling across the city. Among Hùng's expanded duties were disinfecting common surfaces, like the elevator, and performing *quét nhà*, the routine sweeping of the grounds.

This collective cleaning ritual, an everyday urban care practice, extends from homes into the street, transforming the tranquility of daybreak into a cacophony of "swishing" sounds as stiff straw brooms rhythmically brush against the hard concrete surface (audio clip 1). During the pandemic, these auditory cues of shared upkeep devolved into a solitary duty for the security guard, underscoring how lockdown controls reshaped communal routines into atomized activities (figure 11). Throughout the three-week period of "total society quarantine" (*cách ly toàn xã hội*), Hùng bore the dual responsibility of both protecting and maintaining the common areas. This additional, uncompensated care work proved physically and emotionally taxing as he was unable to rotate weekly shifts with his fellow guard who remained stranded in a neighboring province. Separated from his family, he labored day and night under the weight of worry and isolation.

The lockdown did not deprive all affluent families of their household workforce, I would learn. Among expatriate communities in Hanoi, a gendered international division of reproductive labor had long maintained racial and ethnic hierarchies within the care economy (Parreñas 2001). With the closure of schools, many expatriate families returned to their home countries to wait out the pandemic in closer

FIGURE 11. Sweeping common spaces during pandemic time, a familiar morning rhythm of environmental care and shared urban upkeep in Trúc Bạch Ward, 2021. Photo by the author.

AUDIO CLIP 1. Urban caretaking: collective morning sweeping ritual along the contours of Trúc Bạch Lake. Source: author.

To hear this audio clip, scan the QR code with your mobile device or visit DOI: https://doi.org/10.1525/luminos.249.audio1

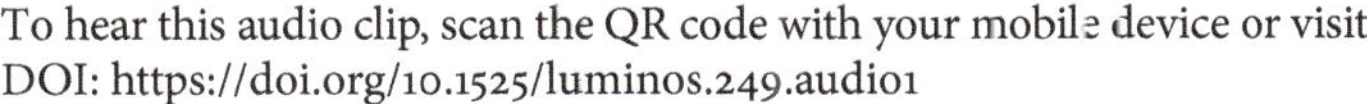

00:00:05   00:00:29

Audio clip 1

proximity to relatives, while others remained in Hanoi. Expatriates were the only group of foreigners I encountered, albeit infrequently, in the streets or markets after the borders closed to tourists and business travelers. Some continued clandestinely employing their household help to offset increased care responsibilities during lockdowns. One such worker was Hiền, a part-time housekeeper in my building who later explained over several conversations how she continued working for a Japanese family throughout the shelter-in-place order, but with reduced hours. This meant shouldering a heavier care workload with elevated risks for less pay. Yet rather than viewing her situation as purely exploitative, Hiền expressed gratitude for having some earnings amid such uncertainty. Her mixed feelings underscored the emotional toll of reconciling survival with the conditions that undervalued her essential contributions.

Hiền was born in rural northern Vietnam in 1968, though her parents altered her birth certificate to 1965 to receive more rations during the war. One of her earliest memories, she vividly recalled, was her father shielding her in his arms amid the thunder of American B-52s, her small body pressed against his chest under his protective straw hat (*mũ rơm*) as they huddled in a trench dug in their home during the air raids in 1972. Enduring poverty and hunger, much of her childhood was spent performing arduous labor, toiling in rice fields or carrying water (*gánh nước*) from village wells to her home and to better-off households for small change. She raised chickens for eggs and pigs that her mother sold in the market to purchase food. These early experiences with precarity would influence Hiền's decision to seek employment in the city. As a young woman, she migrated to Hanoi, where she eventually married but was widowed young and left to raise their son alone. Through an acquaintance, Hiền secured a job washing dishes at one of Hanoi's first Japanese restaurants and steadily advanced to positions with more customer interactions. A quick learner with a gregarious nature and strong work ethic, Hiền impressed the owner, who arranged for her to work as a live-in caregiver in Japan for two years. Although she was well paid, she remembered this time as unbearably lonely; the family was kind but reserved, and people rarely spoke to one another on the street, a stark contrast to the lively chatter and calls of Hanoi's vibrant neighborhoods. Hiền returned to Vietnam with a new domestic skillset and knowledge of Japanese culture, including some of the language. Following an introduction from the restaurant owner, she began working full time as a housekeeper, nanny, and cook for a Japanese family in Hanoi.

The lockdown proved immensely stressful for Hiền, a single mother with substantial care responsibilities of her own. Like many working-class families in Hanoi, her living quarters were cramped—a mere sixteen square meters, shared with her nineteen-year-old son, whose scholarship to study in Japan was disrupted by the coronavirus, forcing him to take courses online from home. Hiền lives in Hanoi's historic Old Quarter, on one of its bustling retail corridors—where the cries of vendors compete with the noise of congested traffic—in a two-story dwelling with her in-laws residing upstairs, requiring her care. When I joked that she occupied some of the city's most valuable real estate, she laughed dismissively, remarking that her dilapidated building was in such poor condition that "no one would consider buying it."

Although her domestic work during lockdown violated official regulations, Hiền maintained that she had chosen this arrangement of her own accord. She framed it as a relationship of mutual necessity, exercising her agency to negotiate terms even while acknowledging the stark inequalities at play. For Hiền, continuing to work was a matter of economic necessity; for her employers, a confined expatriate family with two children, it was about managing the increased burden of household care. With discretionary movement around the city prohibited and public transit suspended, Hiền had to walk the four kilometers to work—a task she had never before undertaken. With the usually cacophonous streets eerily quiet, the echoes of her footsteps threatened to betray her presence. The Old Quarter's maze of narrow alleys gave her an advantage, however, as police checkpoints were mostly stationed at major intersections. Her route also passed several open markets, which provided plausible reasons for being out if stopped. Hiền braved the commute three times weekly, down from her usual five days of work, earning just 50 percent of her usual ten million VND (US$430) monthly salary.

The new labor conditions—and magnified care expectations—proved burdensome for Hiền. "The family was always around," she confided, "and the parents were so stressed." She lost much of her autonomous work ability, constantly under her employer's watchful eye. Gone were the peaceful periods of downtime before the family returned home in the late afternoon. Care demands intensified with the children requiring more of her attention, their voices now filling the home throughout the day. She found herself constantly wiping down surfaces, including floors, to maintain a sterile virus-free environment. And with the entire family now eating all meals at home, amid the constant clatter of dishes and kitchen activity, Hiền felt overwhelmed and unable to complete her tasks to her usual high standards. Her confidence in her work performance faltered.

At day's end, Hiền donned her mask and hurriedly made her way along the deserted streets back to the Old Quarter, stopping at a wet market for fresh produce for her evening meal. The plastic bag clutched tightly in her hand served as evidence to justify her presence on the street should she encounter authorities. Her daily navigation of these challenges highlighted the ways the pandemic intensified gender, class, and racial hierarchies in the informal care economy. As a Vietnamese working-class woman, Hiền managed the triple burden of caring for her expatriate employer's family, her own household, and her elderly in-laws, all while confronting heightened health and legal risks. Her story illustrates how migrant care workers, whose labor was essential to sustaining middle-class families over the course of COVID-19, remained among the most vulnerable and invisibilized members of the pandemic workforce.

Act 2

———

Lockdown

Self-Care

On the afternoon of March 31, 2020, just after my ward-assigned nurse had declared me "virus-free" with a row of smiling emoticons, my phone chimed with a *tin nóng* or "breaking news" notification from state media, followed by a text message from the Ministry of Health announcing, "Việt Nam 'cách ly toàn xã hội' trong 15 ngày" (Vietnam "total society quarantine" for 15 days). I clicked the link, waited for the app to open, and stared in disbelief. The rumor of an impending action was confirmed—a nationwide stay-at-home order would begin at the stroke of midnight on April 1, precisely when my quarantine was set to end. My plans for release would have to wait another two weeks.

The announcement triggered a sequence of orchestrated sound events that shifted my sonic sensibility, while uniting isolated individuals in shared time and space (Ferrarini and Scaldaferri 2020, 1). Authorities swiftly mobilized an assemblage of penetrating, mass communication technologies that collapsed the distinction between public and private spheres to ensure widespread awareness through relational, co-present listening. Mobile broadcasting units, like police sound trucks equipped with megaphones, crawled through the neighborhood (see figure 9), while public announcements reverberated through my ward's loudspeaker system, or *loa phường*, an ostensibly obsolete technology that found renewed sonic relevance during the pandemic.

In re-sounding the public sphere (Lacey 2013, 199), loa phường structured pandemic time through its routine transmission of official communications to an anxiously listening public (as well as a willfully *non*-listening *counter*public).[1] I strived to hone my aural sensibilities and attune my ethnographic ear to actively listen *with*—collectively tuning in to the amplified yet distorted female voice. This voice—not really a "human voice" per se, but as Brian Larkin (2004, 1002) argues, one that has been "transduced, converted from electrical impulses" back into "semi-coherent speech"—competed with acoustic overlays from another neighboring loudspeaker, its soundwaves refracting across the lake to reach my seventh-floor apartment. During the lockdown, these acoustic effects of overlay (dissonance), amplification (volume), and distortion (transduction) constituted my primary urban aesthetic experience of sonic socialism.

The "vocalizations of authority" (Ó Briain 2022, 13) relayed through Hanoi's public loudspeakers—emanating from various government bodies, including the Ministries of Public Security, Health, and Information and Communication, as well as the Municipal People's Committee—largely echoed online and print media content (before its suspension during the stay-at-home order). The prime minister's Directive No. 16/CT-TTg, issued on March 31, 2020, implemented a nationwide policy of containment through social disaggregation. It mandated the closure of all nonessential operations, adhering to a principle of multiscalar isolation: "families isolate from other families, villages from other villages, communes from other communes, districts from other districts, and provinces from other provinces."[2] Businesses trading in essential goods and services (*cơ sở kinh doanh các loại hàng hóa, dịch vụ thiết yếu*) would remain open, and people were permitted to go out for food and medicine as necessary, but only in the immediate vicinity, unless their labor was required at production facilities vital to the economy. Mobility and sociality would be strictly controlled; no private traveling or assembly in groups of more than two people would be allowed. Public transportation, including bus, air, rail, and taxi services, would be suspended until further notice.

As a biopolitical event at the nexus of life, care, and power—carried out in the "spirit of prioritizing people's health and lives above all else" (với tinh thần coi trọng sức khỏe và tính mạng của con người là trên hết)—the preemptive emergency measure aimed to curb the potential for community transmission of the virus. Forecasting the risk of further spread, it came in response to several high-profile outbreaks (*ổ dịch*), including those in Hanoi (in Trúc Bạch Ward and at Bạch Mai Hospital) and in Hồ Chí Minh City (at the so-named Buddha Bar).[3] "Total society quarantine" was "not a national lockdown," government officials emphasized in a broadcast news release issued a few hours later to distinguish Vietnam's more tempered pandemic response from the harsher measures taken by countries facing escalating public health disasters (Viết Tuân 2020).[4] The shelter-in-place order did, however, signal a scaling up of the government's proactive and multipronged "Zero-COVID" strategy, which had swiftly contained localized outbreaks, including in my neighborhood. This strategy encompassed mass testing, targeted quarantine, aggressive contact tracing, universal masking, and the shuttering of schools, bars, and borders. The state's anticipatory approach to pandemic management—responding proactively to the risk of exposure rather than reactively to exhibited symptoms, a defining feature of "disaster socialism," as argued in act 1—kept infection rates low. As of April 1, 2020, when the stay-at-home order took effect, only 218 COVID-19 cases had been recorded nationwide.[5]

While arguably moderate compared to the stricter responses taken in other countries, such as the deployment of military forces in the streets of Lombardy (Italy) to enforce a full-scale lockdown there, Directive 16 was a notch up from the previous Directive 15/CT-TTg, ratified just a few days earlier on March 27 by the prime minister together with the National Steering Committee for COVID-19 Prevention and Control.[6] As a comprehensive set of guidelines of the highest order, Directive 15 called for mobilizing the entire political system, involving all levels of government and the public, from central authorities to the grassroots (*cấp cơ sở*), to expeditiously enact robust pandemic containment measures.

The language used in Directive 15—*đề nghị* (to propose or request)—was firm but suggestive. It also drew on military terminology to convey the gravity of the situation and the need for bold, collective action, including stopping the spread of misinformation. People were "requested" (*đề nghị*) to remain calm, observe spatial distancing measures, and participate in pandemic prevention efforts "voluntarily," or *xung phong tình nguyện*, a term that references historically distinct forms of free labor connected to militia and civilian social services. The Hồ Chí Minh Communist Youth League was tasked with taking an offensive stance (*xung kích*) to inspire voluntarism among its members and encourage public compliance with health guidelines. This was accomplished through social media challenges, like #ONhaLaYeuNuoc (#staying home is patriotic).[7] This challenge was followed a day later by another popular campaign, #ONhaVanVui (#still fun at home), a collaboration between TikTok, UNICEF, and the Ministry of Health in response to Prime Minister Nguyễn Xuân Phúc's declaration that "fighting a pandemic is like fighting the enemy" (*chống dịch như chống giặc*).[8] As a rallying call to action, *chống dịch như chống giặc* worked well as both a politically and an affectively charged soundbite, endlessly repeated in the media but also taken up by artists, songwriters, and social media content creators to rally national support (figure 12).[9] "Every citizen should be a soldier in the fight to stop the pandemic while working collectively and in unity to fend off the disease," Directive 15 urged, explicitly drawing on the metaphor of the virus as an invasive, deadly enemy "on which society wages war" (Sontag 1978, 66).[10]

The language of Directive 16, on the other hand, appeared more forceful, shifting from an urgent appeal to a sober instruction and call for self-discipline, with *yêu cầu*—to require (so as to achieve a particular outcome)—replacing the more suggestive *đề nghị*. Yet even within this sterner order, Directive 16 maintained flexibility in how it framed ethical subjecthood: people were asked to "willingly comply" (*tự giác chấp hành*) with the instruction to shelter in place as part of their civic duty. But would people "willingly comply"? I wondered, reflecting on the growing divide around the world between fear of the virus and skepticism

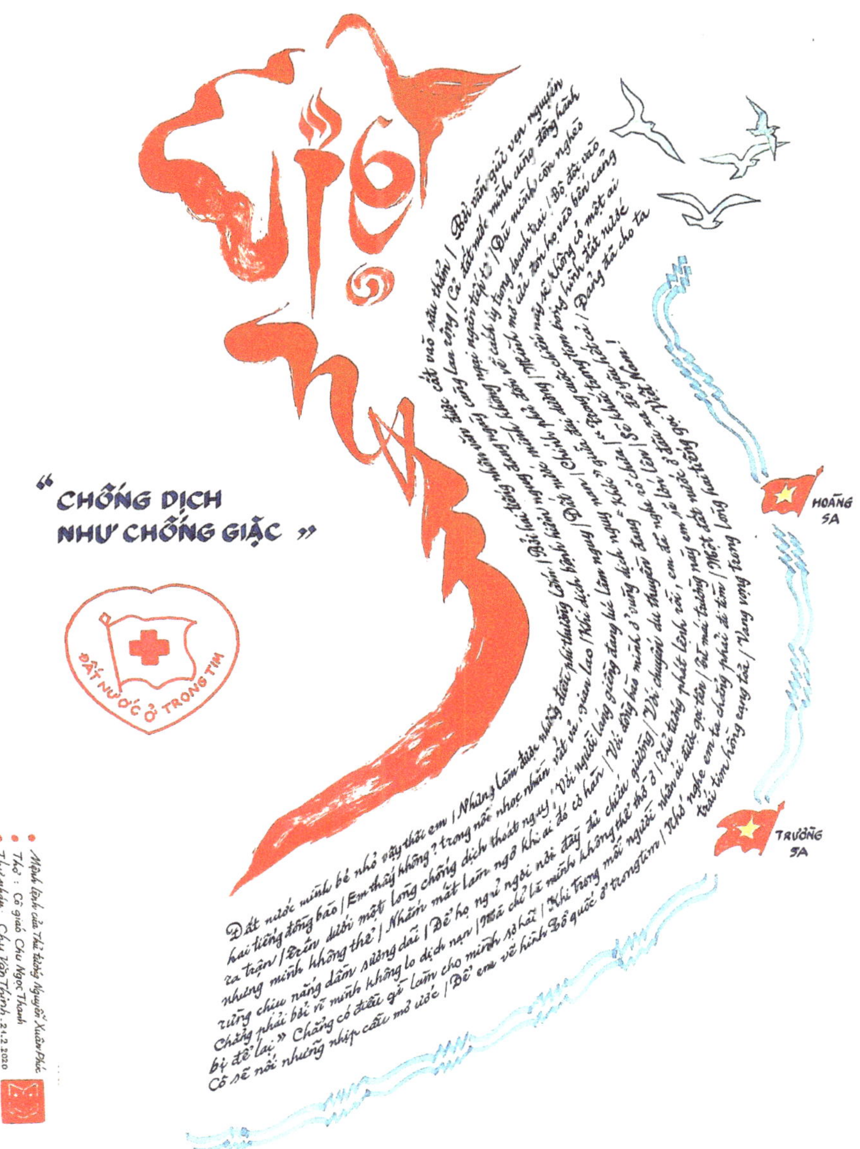

FIGURE 12. "Fighting a Pandemic Is Like Fighting the Enemy." This sound bite took on nationalist overtones, promoting unity while asserting Vietnam's claims to the Trường Sa (Spratly) and Hoàng Sa (Paracel) Islands. Calligraphy by Chu Văn Thịnh, posted on public announcement board in Trúc Bạch Ward, February 2020.

toward measures intended to ensure health security (Charles 2022). In Vietnam, government campaigns and decrees are often ignored or circumvented, until they are strictly enforced. Yet the pandemic proved different; people feared not only the uncertainty surrounding the virus but also the possibility of dying or of causing someone else's death. While the directive mandated staying home, it did allow for essential trips outside, a crucial allowance given my depleted supplies after quarantine.

The next morning, I woke to the loudspeaker broadcasting information about COVID-19's respiratory symptoms, including their sonic identifiers like coughing and heavy breathing. With a shopping bag slung over my shoulder and a mask securely in place, I set out on foot to stock up at the nearby wet market. This brief expedition, my first venture outside in almost two weeks, would expand my soundwork to encompass "footwork" and "walking-notes," adapting my sensory research methods at a time when traditional fieldwork was not possible. Yet my first walk revealed a ghostly stillness, a desaturation of sensory stimulation in this usually bustling city. The lakefront street in front of my building, normally alive with chatty morning walkers and bike riders, stood empty and quiet, no motorbikes or street vendors within sight or earshot, no fragrant *phở* broth wafting through the air, no sweeping sound of straw brooms clearing debris from the curbs.

With my ethnographic sensibilities piqued, I took a right turn and followed the paved road as it wound around the lake, choosing a slight detour over the direct route to the marketplace. The usual team of men playing their pre-work game of *đá cầu*, or foot badminton, with its rhythmic kicking of shuttlecocks, was absent. Missing too were the seniors who roused their bodies on the exercise equipment installed alongside the lake. Silence had replaced the usual morning soundscape of lively fitness activities. Sensing I had gone far enough, I began to loop back around the small island of Ngũ Xã toward the market. On impulse, I crossed the canal (with its persistent smell of sewage) and turned left, continuing along the lake to the "toad market" (*chợ cóc*), where itinerant traders normally gather in early mornings before police chase them away. But they too were gone.

In the distance, I saw a barrier blocking Thanh Niên Road, a major thoroughfare. Realizing I had strayed too far, I began to turn back and locked eyes with a policeman there. He waved his arms and shouted in English: "Go home!"—a command I promptly obeyed. But those brief fifteen minutes outdoors were enough for me to perceive how profoundly the urban atmosphere had changed.

This chapter argues that the nationwide stay-at-home order marked a critical moment of sonic rupture that redefined the city's pandemic soundscape over the subsequent weeks. Acoustic transformation unfolded in distinct phases detailed below, from an ambiance of silence representing the "sound" of compliance to a gradual return of "noise" as expressions of sonic agency that challenged the state's sonic hegemony. This progression reveals how auditory experiences of urban space during crises, along with their associated listening practices, are not only dynamic but also generative of empowering dissonance.

WEEK 1 (PART 1): SOUNDS OF SILENCE

When there is nothing to hear, so much starts to sound. Silence is not the absence of sound but the beginning of listening.
—SALOMÉ VOEGELIN, *LISTENING TO NOISE AND SILENCE: TOWARDS A PHILOSOPHY OF SOUND ART*

Relational Attunements

The pandemic introduced new sensory dimensions to the city, dramatically restructuring the familiar "acoustical composition" of urban life (Farina 2014). Around the world, lockdowns quieted the usual cacophony as cities came to a standstill. Hanoi, a sonically vibrant city with minimal regulation of noise, was no exception. The pulsating, sonorous qualities that defined its sonic identity and the aural intensity of its streets disappeared. Absent, too, were the everyday auditory experiences that constituted Trúc Bạch's lively soundscape: the hammering from construction sites, the honking of traffic, the hum of motorbikes, the clamor of pupils at recess, the chopping of herbs at soup stalls, the tapping of wooden tocsins (*mõ*) in pagodas, the tolling of church bells, the signal "clang-clang-clang" of waste collectors, the offkey karaoke singer, and the cries of itinerant street traders ("Who wants to buy sandals?" [Ai mua dép giày?]), one of the Hanoi's most iconic soundmarks. Even nonhuman vocalizations that contribute to soundscape ecologies—the barks of dogs, trills of birds, and crows of cocks—seemed to vanish. As people left town or retreated behind shut doors and secured alleyways, the usual din of domestic life spilling into public spaces faded, leaving the city silent and deserted.

As certain sounds faded, others came to the fore. Silence alters our sonic sensibilities, Salomé Voegelin observed, heightening our capacity to perceive by shifting attention to the less detectable hums, pings, murmurs, and whirrs that are more often sensed—felt through the body—than heard. Voegelin writes of the "persistent pulse" of the dripping icicle on her balcony, each drop exuding a different aural quality (2010, 85). "This is not really hearing but sensing sound," she describes of the visceral experience. "Sounds are tangible in this dense quietness. I am feeling through my body whole clumps of sensate material. The quietness enhances my perception; I take notice of every whisper, hum and buzz" (2010, 85).

Similarly, silence afforded me the opportunity to attune my senses to the nuanced sensations around me while reflecting on the "enculturated nature" of urban sounds (Samuels et al. 2010, 330). On that first evening of the shelter-in-place order, I stood on my balcony, engulfed by an unusual stillness. As I listened intently to the silence, I began to hear—and indeed, to feel—the subtle murmurs of urban infrastructure: the low vibrations of electricity, the faint whir of current from the nearby power grid, and the buzzing of tangled transmission lines. I detected the rhythmic mechanical pumping of water through pipes to rooftop tanks. In the distance, the rumble of a garbage truck compacting its haul momentarily broke the subdued quiet. This sound evoked in me thoughts of frontline workers in critical industries whose undervalued labor and care work, performed in the shadow of the pandemic, sustained essential operations across the city—including those that I was sensing that evening.[11]

The lockdown's stillness amplified the ambient sounds of nature, revealing the more-than-human geographies that are integral to the sonic ecology of the mediated city (Atkinson 2007). Dwelling in urban silence compelled an "atmospheric attunement" of my senses to broader lifeworlds (Stewart 2011), enabling a deeper, more perceptive practice of relational listening, or what Steven Feld (2015, 15) has called "modes of acoustic attending" to co-presence. The cadenced croaking of frogs, energetic chirping of birds, shrill chorus of cicadas, faint whistling of wind, and gentle lapping of water all gained prominence in the absence of the laughter, toasts (*Một, hai, ba, dô!*), singing, and socializing that once echoed nightly from the outdoor cafés, beer joints, and hotpot stalls lining the lake's perimeter.

The Deserted City

The tranquility that settled over the city was not only an effect of families self-quarantining and working from home behind shuttered doors. It was also an outcome of urban flight. The stay-at-home order emptied the city as large numbers of people—urban residents and migrants facing economic uncertainty—fled to their home villages before the travel ban went into effect.[12] Even those essential businesses permitted to remain open closed up shop and left for rural provinces. In my neighborhood, the barbers, café workers, shoe shiners, taxi drivers, market vendors, and fruit sellers—along with the owners and employees of corner stores, eateries, clothing boutiques, repair shops, and pharmacies, and even my land-lord—packed up and left, many with their families (and pets), before movement across administrative borders was restricted.[13] Some wards, like neighboring Yên Phụ peninsula with its single-lane entry point, self-organized to blockade outsider access to their streets, also curbing mobility within Hanoi as a grassroots strategy to reduce risk through self-containment.[14]

The idea of "the rural" as safer ground than "the urban" has long shaped patterns of mobility between city and countryside, also in wartime.[15] As Frank Proschan (2002, 620) has argued, "The emergent colonial cities of Hanoi and Saigon, Haiphong and Danang came to represent hotbeds of cosmopolitan degeneracy" and racialized disease, in contrast to the "idyllic and timeless countryside of the traditional peasantry." Beyond imaginaries of rural peace and purity as an escape from COVID-19, there were practical reasons for the urban exodus before the lockdown.[16] My interlocutors described advantages such as family reunions, improved material conditions (particularly for migrants, who typically shared cramped dormitory rooms in old buildings), food security, lower costs of living, and more relaxed pandemic regulations.[17] One of my neighbors, who had returned to the city alone to reopen his hair salon weeks later, explained that his wife and children were better off with his parents in the countryside, beyond the reach of the virus. The rural setting provided ample space and a modest livelihood from the family's poultry and paddy crop. They also had help with childcare, and his sons could play outdoors in the fresh air surrounded by gardens and fruit trees. It was indeed a bucolic image of the good life, one that has increasingly lured young urbanites to "opt out" of their careers in noisy, polluted cities and embrace pastoral life through farming and agribusiness activities in more tranquil and healthful surroundings.[18]

The Time of Silence

Urban tranquility during the lockdown had a distinct aesthetic temporality. Take nocturnal time, for example.[19] The flight to the countryside left much of the architectural fabric in my neighborhood eerily dark and still at night under the dim, yellow glow of streetlights, from shuttered homes to shops and hotels. This urban darkness evoked an uncanny sense of impending apocalypse, with ghostly landscapes altering both the visual and sonic dimensions of a city that felt vacant and suspended in time. This perception was not unique to Hanoi. Online articles—and now, books (see Loria and Loria 2021)—have highlighted the uneasy urban hush that fell over much of the world, yet often with little attention to the specific historical and cultural contexts that shaped how pandemic politics and crisis management unfolded differently across the globe.

Likewise, much has been written about the relationship between time and the pandemic, particularly its impact on emotional health, such as increased levels of anxiety and an acute feeling of temporal uncertainty. Across the globe, people reported experiencing time as distorted, though how they understood and attributed meaning to these experiences remains underexplored. There was a pervasive sense of losing track of time while sheltering in place, as everyday routines were disrupted. Some felt time sped up and stretched out, while others felt it slowed down and compressed. Still others felt stuck in time, trapped in an endlessly liminal present, with days running together, a sentiment that gave rise to the term "Blursday" (Williams 2020, Gupta 2022).

The loss of awareness of time has been linked to disruptions in "temporal landmarks," or regularly scheduled activities associated with particular spaces or habits that structure daily lives (Grondin, Mendoza-Duran, and Rioux 2020). Beyond individual calendars, our temporal orientations are also informed by the ability to sense wider atmospheric changes, such as seasonal transitions or daily fluctuations in the tones and intensity of light. Essential to my argument about sonic techniques of pandemic governance is how our perception of time is punctuated by sound—and its absence. Auditory cues orient our days and help us track time: the school recess bell (its silence signaling the weekend), the delivery thud of daily mail, or the chorus of birdsong at sunrise. This temporal marking through sound is not new. Alain Corbin (1998) famously documented how village bells in nineteenth-century France served as acoustic instruments of power, announcing the passage of time while fostering a connection to place. The sounds and technologies introduced by industrialization—like the punch clock—would later displace bells as timekeepers, radically altering the rhythms of modern life.

Changes to soundscapes during COVID-19 similarly altered temporal perceptions. Yet discussions of pandemic temporalities have largely overlooked how sonic dynamics affected our sense of time and space.[20] Awareness of time is shaped not only by familiar routines—upended by COVID-19—but also by familiar sounds, many of which ceased or diminished during lockdowns. The widespread experience of pandemic timelessness, therefore, cannot be attributed solely to the absence of temporal landmarks. It also arises from the loss of critical "soundmarks," those distinctive auditory features of our shared acoustic environment that collectively shape our relationships to time (Schafer 1994, 10).

WEEK 1 (PART 2): BREAKING THE SILENCE

Loudspeakers are the power of the regime and the thread that binds the government and the people.
—GENERAL BẠCH THÀNH ĐỊNH, DEPUTY DIRECTOR OF HANOI PUBLIC SECURITY

Sounding Discipline: Urban Loudspeakers as Panaudicon

The stillness that swept over the city left a void for other public sounds to fill. From the silence of the lockdown, an iconic yet seemingly outmoded broadcast technology was resurrected: the public loudspeaker, or *loa phường*.[21] A fixture in every ward, the public loudspeaker stands as a quintessential symbol of Vietnamese state and party power intended to create "a sense of proximity between state and society" (Ó Briain 2022, 87). Similar to quarantine (act 1), during lockdown, when direct supervision proved difficult, the state asserted its presence through the tactical deployment of media technologies—tools integral to its governance of care. Loudspeakers in particular emerged at the center of an "acoustics of power" that functioned as a Panaudicon, through which the state's authoritative knowledge would be both felt and known through immersive sound (Rice 2003, 7).

Like the radio, the loudspeaker functions as an apparatus for one-way distribution (Brecht [1932] 1967, 134), enhancing its sonic authority by redefining the technological interaction between a speaking object and a listening subject. As part of an extensive communication infrastructure, loudspeakers enabled the local People's Committee to strategically amplify its voice, projecting expert-driven knowledge loudly and instantaneously to create an inescapable "panauditory omnipresence" (Siisiäinen 2013, 58). The twice-daily public announcements in my ward—at 6 a.m. and 4:30 p.m., each lasting around thirty minutes—functioned as temporal soundmarks during the timelessness of lockdown.[22] The resounding female voice, projecting a present stable state, cut through the stillness of dawn, signaling the start of the day. The hailer's proximity to my bedroom window made

FIGURE 13. Loudspeaker attached to utility pole next to power station, Trúc Bạch Ward, 2020. Photo by the author.

it impossible to ignore. Indeed, the loudspeaker's ability to "impinge" (Khan 2011, 580) on private space and permeate the city's acoustic atmosphere—despite widespread inattention, especially among youth—made its pandemic revival controversial, particularly in wards that opted to broadcast throughout the day.[23]

Urban loudspeakers have long been at the center of public debate regarding their ongoing use as a mode of public address in the digital era of e-governance. Many Hanoians have expressed an ambivalent relationship to the loudspeaker system, which is deeply embedded in the city's architecture: strapped to utility poles, affixed to trees, or placed atop cultural centers or other administrative buildings, a pattern common across much of urban Vietnam (figures 13 and 14). Before the pandemic, many of the loudspeakers in Hanoi's inner-city wards had been deactivated, including in my neighborhood, where tree canopies had engulfed the city's acoustic infrastructure under their foliage. A 2017 e-government survey found

FIGURE 14. Loudspeakers mounted on cultural center in Quang Trung housing complex in Vinh City, designed to project sound in multiple directions, 2011. Photo by the author.

that 90 percent of respondents, notably those with the digital skills to navigate e-survey platforms, advocated for abolishing the loudspeaker system, pointing to the rise of new media technologies that rendered their intrusion unnecessary (Vo Hai 2017).[24] Citing a digital divide and unequal access to information technology, authorities rejected the survey's findings, arguing that loudspeakers continue to be an essential tool of governance, particularly in peripheral wards and communes with limited digital connectivity. They did, however, agree to restrict their use in central urban districts to a public warning system, rather than a "routine sonic accompaniment" to everyday life (Ó Briain 2022, 13).

The Wuhan outbreak catalyzed the reactivation of loa phường, bolstering the state's position that loudspeakers remain a vital mass media technology for the city's emergency response. This "resurrection" (*hồi sinh*), as dubbed in the press, helped wired loudspeakers shed their perceived "backwardness" (*lạc hậu*) and fulfill their "lofty mission" (*sứ mệnh cao cả*) to disseminate urgent public messages quickly and universally, reaching "every person and household" in the city (*đến mọi người mọi nhà*) without regard to social hierarchy (Quỳnh Hoa 2020).[25] In the logic of the state, loa phường promised a more accurate and dem-ocratic medium of mass communication that hailed all listeners, including those from more vulnerable groups, as citizens of an inclusive sound community. They bridged distances and overcame divisions reinforced by digital technologies. Their "compulsory egalitarianism" (Coderre 2021, 38), which subjected everyone to identical aural stimuli, discouraged individualized listening, while reinforcing socialist claims to collectivity and cohesion in alignment with broader state and party objectives.[26]

Red Noise: Sonic Socialism

The joy of every home resounds through the loudspeakers
Sweet spoken voices and enthusiastic song . . .
—KIM CÚC, "LISTEN TO THE VOICE OF THE HOMELAND"

As the state's go-to sonic technology to manage pandemic uncertainty, loa phường underscored the historical significance of oral public spheres in urban mobilization campaigns and the cultivation of a communal—and communist—ethos in Vietnam. This ethos embodies what I have termed in this book "sonic socialism," where sounds played a pivotal role in forging emotional bonds and appropriate forms of relational personhood to advance socialist nation-building.[27] Socialist sounding—a process of "making and imagining" that strives to civilize and territorialize through sonic tools (Coderre 2021, 32, 52)—demonstrates that sounds operate as active, politically charged agents of revolutionary change rather than passive backdrops to lived environments.

While urban histories—including recent pandemic narratives—often privilege sight over sound and panopticons over panauticons, the sonic dimensions of landscapes warrant closer attention. As Shannon Mattern (2015, 98) observes, media infrastructures shape urban morphologies beyond just print and visual cultures. The voice itself has been "built into urban form" and inscribed onto city surfaces, with acoustic considerations influencing design and construction since the earliest cities, serving both social and governance functions (2015, 98–99). This reveals cities as complex, mediated sensory environments, where humans not only develop and deploy media technologies, but are themselves "already technologically mediated" (Larkin 2014, 990).

In Vietnam specifically, sound production and its architectures are deeply entwined with the development of colonial and postcolonial cities. This auditory history parallels the evolution of sound-reproduction technologies, including public loudspeakers, which transformed urban environments and sensory experiences. Broadcast media proved critical to decolonization and state formation, reaching a largely illiterate population through the rapid dissemination of carefully scripted news, public service information, and cultural programming designed for universal audibility.[28] From makeshift megaphones used in the absence of electricity to mobile broadcasting vehicles serving as radio "studios on wheels" and wired loudspeaker systems akin to Soviet "spoken newspapers" of the 1920s (Lovell 2015, 2), this emergent sonic geography became imperative to the revolution and its efforts to dismantle the colonial regime (Ó Briain 2022, 71).[29] Sonic socialism operated on multiple levels, facilitating mass communication and political socialization while cultivating collective identity through shared auditory experiences, transforming a listening public into an enlightened and ideologically aligned population.

The revolution's sonic infrastructure laid the groundwork for the postcolonial state's communication strategies, particularly in disaster preparedness. Hanoi authorities have emphasized the public address system's dual role in the capital's history: building a "cultured and civilized" way of life while safeguarding public order and national security. During the war with the United States, broadcast systems proved indispensable, as loudspeakers alerted citizens to impending air attacks (Nguyen-Thu 2020, 145). A decade of aerial bombardment fostered reliance on one's auditory acuity for survival. My research respondents, for example, described how they developed heightened sensitivity to sonic indicators that signaled imminent danger or its abatement while seeking refuge underground (Schwenkel 2020, 64). An elaborate alarm system—incorporating sirens, drums, gongs, and bells, some fashioned from bomb casings—kept people informed and forewarned. This network of acoustic devices, from rudimentary to advanced, enabled the transmission of instantaneous warnings. Public loudspeakers amplified this sonic network, issuing urgent public orders and emergency communications, such as mandatory evacuations of the city.

During the postwar era, loudspeakers remained a cornerstone technology in the sonic governance of Vietnamese cities. Materially intertwined with urban design and affectively tied to hardships of the subsidy period (through 1986), they served as a political tool for organizing and rebuilding society. While routine broadcasts of announcements, news, events, and musical entertainment made loa phường a regular acoustic presence in everyday life, sound artist Margarethe Maierhofer-Lischka (2024) observes that what emanates from loudspeakers is not always related to what is otherwise audible. This gap between acoustic transduction and sensory perception—between sound produced and sound heard—reveals how loudspeakers transcend their role as mere technical objects for sonic dissemination; they are cultural and political artifacts whose historical significance generates powerful affective resonances.

This disjuncture between the technological (sound as objective) and the experiential (sound as interpretive) underscores the political stakes for the future of loudspeakers as a medium of public address. Discussions of loa phường as disruptive, sonic intrusions needing deactivation often overlook class and generational divisions, as well as who has the social capital to navigate digital environments. Criticism of loudspeakers also tends to ignore the more positive associations they hold for different social groups, particularly for older generations who may lack technological fluency but associate these devices with wartime affects of pride and solidarity.

Hanoi officials have framed this debate in terms of privilege, attributing loudspeaker aversion to affluent, digitally literate youth who like to "sleep late," implying a lack of work ethic found among older, more appreciative listeners. For this latter group, loudspeakers evoked a visceral sense of nostalgia for a difficult past, what I have called "Vinastalgia" to capture the particular socialist affects and emotional resonances of objects from that period (Schwenkel 2022a). Among these "nostalgic listeners" (Ó Briain 2022, 112), vintage loa phường have attained the

FIGURE 15. Retro sonic artifacts: wartime audio heritage on display at Symbiotic Café in Hanoi, 2022. Photo by the author.

status of commemorative objects, now displayed in history museums and subsidy-era-themed cafés, where they serve as focal points for sonic sociality (figure 15).

The arts scene in Hanoi has shown particular fascination with these retro audio icons and their symbolic significance. In August 2022, misunderstood loudspeakers became the subject of a performance piece where installation artist Yến Năng "argued" (cãi) with a mute speaker—a proxy for the state—about its fate, pushing for its conversion from a tool of one-way transmission to a medium for two-way communication, as Brecht ([1932] 1967) envisioned of radio. Offering an alternative perspective, Nguyễn Thế Sơn's audiovisual exhibition *Sắc màu của âm thanh* (Colors of Sound) suggested that public loudspeakers and their sonic ecosystem—comprising poles, wires, trees, speakers, and the sounds they emit—constitute urban heritage worthy of preservation (figure 16). For these artists, loa phường are not mere sensory assailants demanding to be heard but rather chroniclers of Hanoi's complex urban history of social transformation. Through their commanding aural (and material) presence, loudspeakers emerge as active guardians of historical memory, refusing to let the socialist government's sonic responses to past catastrophes fade into inactive silence.

These early deployments of sound reproduction technologies in anticolonial and anti-imperialist struggles coincided with a period of heightened party legitimacy

FIGURE 16. *The Red Days III* (Những ngày đỏ III), silk painting by Nguyễn Thế Sơn, 2017. Courtesy of Nguyễn Thế Sơn.

FIGURE 17. Solidarity stamps commemorating daily medical interventions and technologies with the slogan "Join Hands to Prevent and Control COVID-19." Issued by the Ministry of Information and Communications on March 31, 2020, to honor the *chiến sĩ áo trắng* (soldiers in white dress), a historic term for battlefield medical teams adapted to recognize health care workers during the pandemic.

in Hanoi, where the socialist state used sound to consolidate its power and assume the role of paternal protector during national crises. Nearly five decades later, this historical precedent provided the rationale for resuming loudspeaker broadcasts in the government's new "war" against COVID-19, invoking battlefield rhetoric to galvanize public sentiment. Echoing past conflicts, this encounter with an invading "enemy" (the coronavirus) was framed and "fought" acoustically as a sonic campaign to rally the population against what was presented as yet another external threat to the survival of the Vietnamese nation-state.

Sonic Campaigns of COVID-19

During the April 2020 lockdown, as people were confined to their homes, sound media became one of the state's most potent tool of mass communication to ensure the protection of public health. This meant that the primary mode for cultivating "moral citizenship" at the time was not visual (Bayly 2020), but auditory. While pandemic iconography existed (figure 17), auditory technologies achieved greater cultural penetration.[30] The resonant voices—predominantly female, both live and recorded—transcended material and digital barriers, their vibrations reverberating through walls to deliver real-time "essential information" (*thông tin thiết yếu*) to a homebound population in ways that state billboards or posters could not. As conduits of authoritative expertise, these sounds interpellated listeners as moral-hygienic subjects by appealing to a collective sensibility of duty and consent. Just as Foucault observes that the seventeenth-century plague was "met by order" through town-wide quarantines, so was the SARS-CoV-2 pandemic met by a *sonic* order that similarly functioned to "sort out every possible confusion" between body, place, death, and disease (1977, 197). Yet, in this case, disciplinary power manifested not through the omnipresent gaze of the state, but through acts of pervasive sounding to establish acoustic authority over urban space.

A pivotal shift in this sonic order, from preemptive preparedness to an all-out emergency response, occurred on March 8, two days after the discovery of Patient 17 put a street block in Trúc Bạch Ward under collective quarantine (see pandemic figure 1). On that day, Vũ Đức Đam, deputy prime minister and chair of the COVID-19 National Steering Committee, announced the start of a "new campaign" (*chiến dịch mới*) in the phase 2 "battle against COVID-19" (*cuộc chiến chống* COVID-19). Phase 2 called for utilizing all available means to strengthen government coordination to mobilize the "entire population to fight against an enemy" (*toàn dân chống địch*) now feared to be lurking inside the country.[31] To counter this "silent ambush" (*âm thầm mai phục*), the new campaign would launch a full-scale offensive using mass media and information technologies to "equip each person with accurate and accessible knowledge for any location or situation," he announced while standing before a whiteboard sketched with virus transmission calculations (Trần Mạnh and Đình Nam 2020). Acoustic techniques, in particular, would prove essential to "combatting" a virus imperceptible to the human eye and ear, by leveraging the power of sound.[32]

In his work on the auditory atmospheres of warzones, J. Martin Daughtry proposes the concept of "sonic campaigns" to describe the dynamic and often violent relationship between sounding and listening as emplaced and embodied experience within "inherently combative soundscapes" (2015, 127). Inspired by the militarized discourse used by Vietnamese authorities during the pandemic, I extend Daughtry's concept beyond the modern battlefield to encompass other crises that impact sensory modalities, including audition. Sonic campaigns can be understood as deliberate tactics designed to influence conflict outcomes—in my case, "victory" (*chiến thắng*) over the virus—through strategic incursions of sound. While these interventions exert power in ways that may reveal their potential for harm (2015, 126), they can also provoke resistance, disrupting intended effects.

Though operating in a less violent context than the sonic campaigns of the Iraq War that Daughtry examines—where sounds functioned as weapons of terror, recalling the tactics used by Americans on Vietnamese battlefields—Vietnam's COVID-19 sonic campaigns similarly evolved alongside the pandemic's progression, adapting governance strategies to contain an unpredictable and mutating virus. The "intrusion" (*xâm nhập*) of a dangerous, border-penetrating pathogen necessitated the activation of the state's arsenal of sounding practices, aimed at "raising awareness of responsibility" (*nâng cao ý thức trách nhiệm*) across all segments of an increasingly stratified society.

During the period of social isolation—a time of otherwise uncanny urban stillness—state health care mandates broadcast by public address systems dominated the city's soundscape, crafting authoritative messages to reach an "imagined listening community" (Birdsall 2012, 109). These broadcasts disseminated expert-driven information as medical fact, working to combat disinformation while cultivating rational listeners committed to preventive self-care. Through this acoustic regime, the state positioned itself as protector of the population against both an external enemy (the virus) and an internal adversary (fake news), with those spreading "misinformation" subject to legal penalties. In an appeal to self-regulation, loudspeaker scripts transmitted state-sanctioned, evidence-based guidelines for the containment of COVID-19, promoting everyday "good hygiene practices" that bridged science and sensibility, beginning in the home:

> To proactively prevent respiratory infections caused by coronavirus, the Ministry of Health advises people to implement the following health measures: maintain personal hygiene; wash hands with soap and clean water; cover mouths when coughing or sneezing; wear masks in public places; exercise; eat cooked foods and drink boiled water; consume adequate nutrients; regularly clean floors, handrails, stairwells, and doorknobs; sanitize objects with bleach; increase ventilation in rooms; avoid contact with people who exhibit signs of illness; individuals with suspected symptoms should refrain from socializing or visiting crowded places.[33]

As a respiratory illness that produced distinct audible symptoms, Hanoi residents were advised to *listen* and *feel*, rather than *look* for signs. "Seek immediate medical attention if [you] develop a fever, cough, or difficulty breathing," a disembodied voice instructed me. Acute sensory attunement to "sonographic events" (Brown et al. 2020), like coughing or wheezing, could inhibit "severe disease progression that might lead to pneumonia or respiratory failure," it warned with finality, heightening my fear of the virus. In attuning to audio cues as diagnostic indicators of disease, sounds both marked and stigmatized bodies as carriers of COVID-19.

Loudspeakers served as a sonic channel for communication across scales of government, from the COVID-19 National Steering Committee to municipal departments, districts and wards. The nature of information being broadcast influenced the format and quality of sounds produced, whether live or pre-recorded, with inaudible segments sometimes needing to be redone. At the national level, the Steering Committee released updates on countrywide infection rates and disease control mandates twice daily at 6 a.m. and 6 p.m. This structure allowed urgent dispatches (*công điện khẩn*), such as the stay-at-home order under Directive 16, to be transmitted promptly to residents though live broadcasts, with the system's panauditory reach ensuring near-universal access to impactful breaking news (*tin nóng*).

AUDIO CLIP 2. Ward affairs through *loa phường*: female broadcaster hesitates during live reading of health screening notice for military conscripts. Source: author.

To hear this audio clip, scan the QR code with your mobile device or visit DOI: https://doi.org/10.1525/luminos.249.audio2

Standardized, pre-recorded broadcasts, on the other hand, with information distributed to wards by the Department of Information and Communications, in collaboration with local People's Committees, conveyed city-level policy updates and public health precautions, aiming to discipline listeners through reiteration. These broadcasts kept the public informed about critical changes to municipal procedures. For instance, after an outbreak in Đà Nẵng on July 25, 2020—during peak tourist season and following ninety-nine days without recorded community transmission—Hanoi's public address system delivered an emergency citywide directive. This reiterated message, which blended traditional broadcast and digital outreach, instructed returning vacationers to submit medical declarations on the Ministry of Health website, self-quarantine at home for fourteen days, and obtain a COVID-19 test at their local health station.

Beyond this top-down communication, loudspeakers functioned as spoken notice boards, or *bảng thông báo*, which are ubiquitous in urban wards.[34] Even the term used for public address systems, "loa phường," embeds the sound transducer (*loa*) within the jurisdiction of the ward (*phường*), the lowest administrative unit of government vested with authority to disseminate information in alignment with official decrees. As such, wards maintained a degree of autonomy in shaping content and programming, tailoring live and pre-recorded broadcasts to include neighborhood news, local events, and even musical preferences (audio clip 2). Consequently, sound media were just as capable as print media of fostering a sense of shared—if not even imagined—identity.

During the pandemic, neighborhood sonic campaigns in my ward helped translate national policy into local action. Announcements reported infection numbers in the city, outlined procedures for COVID-19 testing, and issued targeted appeals. For example, messages called on residents who had traveled on specific flights or frequented establishments with confirmed cases to come forward and declare their current health status.[35] They also announced the closure of nonessential businesses specific to the area: "Effective from noon today, certain service establishments will be temporarily closed to prevent the spread of COVID-19 until further notice from the city. Specifically, dine-in food operations will be suspended; only take-out or delivery will be allowed. Gatherings with large numbers of people eating and drinking are prohibited. Barbershops and hairdressers must close as well."

Frequently, these broadcasts, prepared by ward authorities and often read in real-time, stirred national pride by ending with revolutionary music intended to uplift spirits. One "red song" that resonated with older generations in particular, "Tiếng Chày Trên Sóc Bom Bo" ("The Sound of the Pestle in Sóc Bom Bo"), celebrated solidarity in the face of adversity.[36] To engage a postwar generation less inspired by red music, broadcasts introduced the upbeat tune "Việt Nam Ơi!" (Hey Vietnam!), a soccer cheer turned COVID-19 anthem and superhero video produced by the Ministry of Health (see act 3).

The emphasis on "grassroots communication" (*thông tin cơ sở*) to keep all social groups informed about ward-specific matters, such as vaccine rollouts, made loudspeakers relevant again in daily life. While digitally-connected youth often criticized loa phường and developed counter-practices to cope with the perceived noise, public broadcasts nonetheless compelled different modes of listening, whether attuned to or tuned out of an acoustically mediated environment. This interaction reflects a broader historical pattern: modernity transformed auditory worlds by introducing new acoustic technologies, such as loudspeakers, which altered how sound interacts with space and how people engage with sonic objects. At the same time, these changes gave shape to modern sonic sensibilities, requiring more refined listening practices (Thompson 2002, 118). In the West, listening as learned cultural practice became deeply entwined with a politics of class and racial difference that reinforced "sonic color lines" (Stoever 2016) and boundaries between colonizer and colonized (Ochoa Gautier 2014). Who listens, how, and to what we attune our ears raise critical questions about power, privilege, and the voices considered worthy of being heard.

In Vietnam, socialist notions of modern personhood are closely tied to state-endorsed listening practices, which are historically grounded in experiences of collective listening that span—and reinforce—social divisions along gendered and generational lines.[37] In my neighborhood, groups of retired men, many of them war veterans, gathered early mornings to discuss current events, tuning in to broadcasts on transistor radios while sipping cups of hot green tea from female vendors along the lake. Meanwhile, women listened to loudspeaker announcements while engaging in morning activities like exercising or managing household duties. This capacity to multitask—listening while working—makes loudspeakers particularly suited to modern life, argued a 2022 Voice of Vietnam radio broadcast titled "Hanoians Still Want Loudspeakers, But Just in a Different Way," which defended their relevance against claims of obsolescence.[38] The broadcast featured a retired woman in Hai Bà Trưng District who credited loudspeakers with boosting her (unpaid) labor productivity during the pandemic, enabling her to care for her grandchildren and manage domestic tasks while staying informed about the time and place of her next vaccination. Yet this example reveals a surprising irony: loudspeakers, once envisioned as a sonic tool for achieving a more equitable socialist future—including gender equality—now inadvertently facilitated the transfer of care responsibilities from the state to households, particularly to grandmothers. In this way, public loudspeakers during the pandemic reproduced a gendered and generational division of labor, supporting an increasingly neoliberal structure within modern Vietnamese families, while placing greater caregiving burdens on women, especially older female kin (see also pandemic figure 3).

Sonic Convergences

Debates over the relevance of Hanoi's loudspeakers in a Silicon Age—whether they align with the ways people listen and access information—frame media technologies as static across time, with rigid boundaries between analog and digital systems. However, this perspective overlooks the dynamic nature of media ecologies. Online and offline modes of communication do not simply supplant one another; they converge and interpenetrate in response to evolving urban conditions, including declared public health emergencies. As "old" infrastructures "leak" into new media landscapes (Mattern 2015, 103–4), they become deeply networked into the spatio-material fabric of the city, creating hybrid systems that build upon residual technological scaffolding (Chadwick 2017).

With an eye toward interconnected rather than independent systems, these "media convergences" (Meikle and Young 2012) alter and remediate traditional technologies through emerging media, rather than rendering them obsolete. Similarly, a *sonic* convergence of media technologies was evident in Hanoi's public address system, which government officials both revived and remediated in response to the pandemic. As authorities became ever more reliant on sound media to communicate health directives, they innovated by integrating digital technologies into the hard-wired soundscapes of sonic socialism, dissolving binary distinctions between old and new. This "remediation" did not entirely absorb or erase the residual medium of the loudspeaker but instead ensured its repurposing and preservation (Bolter and Grusin 1999, 47), resulting in multiple "instantiations of a shared infrastructural purpose" (Mattern 2015, 108).

Remediation as innovation reduced the labor of live transmissions while increasing broadcast time and frequency. Several times a week, women officials from the ward's cultural center in my neighborhood would roll a battery-powered, portable subwoofer out onto the sidewalk. Positioning it in front of the building, beneath the wired loudspeaker, they would insert a memory stick into the USB port and hit play (figure 18). The looped audio file, which repeated the scientific mantra of sensory warning signs alongside other hygiene guidelines—beware of "*sốt, ho, khó thở*" (fever, cough, difficulty breathing)—produced a distinctive masculine vocal aesthetic, marking a shift from the loudspeaker's female gendered voice. The accelerated speech pattern and tonal inflection of the AI-generated narration echoed the male commercial voiceovers heard outside business establishments across Hanoi. From my seventh-floor balcony, I listened to the neighborhood's response to this looping track in the days leading up to the lockdown. At times,

FIGURE 18. Portable Sansui speaker with retractable handle as supplemental media technology: bridging the gap in public health communication between loudspeaker broadcasts, Trúc Bạch Ward, 2020. Photo by the author.

AUDIO CLIP 3. Sonic convergences: amplified state messages via portable subwoofer compete with ambient sounds. Source: author.

To hear this audio clip, scan the QR code with your mobile device or visit DOI: https://doi.org/10.1525/luminos.249.audio3

00:00:01   00:01:04

Audio clip 3

café staff would crank up their own music to drown out the recording, demonstrating how acoustic dissonance could serve as a productive catalyst, not merely a disruption (audio clip 3). By choosing which sounds to amplify or ignore, listeners asserted sonic agency over their auditory environment rather than passively accepting state-designed soundscapes and their messages.

Not everyone ascribed to the moral and political virtues of listening during the pandemic. The sonic agency of baristas exemplified how some residents resisted attuning their actions to the "sermons" of the state, despite political and ethical incitements to comply (Hirschkind 2006, 9). Public address systems, in all their broadcast modalities, made mass communication events so regular a sonic feature of the landscape that they frequently sparked inattention, while others kept an ear open for useful information. If sounding was an exercise of state power meant to mediate and control the city during crises, then diverse listening practices—from attentive engagement to deliberate disregard—and modes of noisemaking offered spaces for subtle defiance. Attuning to these acoustic expressions during the second week of the lockdown, when people increasingly ventured outside, reveals that residents were not uncritical listeners caught between state monitoring and self-regulation. Instead, their listening practices positioned them as "active social participants" (Samuels et al. 2010, 335) engaged in acts of counter-sounding within and beyond the dominant acoustic architectures of urban space.

WEEK 2: RETURN TO NOISE

Today, every noise evokes an image of subversion.
—JACQUES ATTALI, *NOISE: THE POLITICAL ECONOMY OF MUSIC*

Performing Compliance

If week one of the lockdown was marked by an uneasy silence, punctuated by amplified public health directives, week two saw a gradual resurgence of urban noise, as sonic agents found creative ways to circumvent those very regulations.

Urban silence signaled compliance with COVID-19 measures and public trust that the government was acting in the best interests of the people to protect collective health. At this moment in the pandemic, my neighbors not only voluntarily complied with the regulations, but they expressed high levels of public confidence in authorities. This local sentiment aligned with broader trends: A survey conducted by the Mekong Development Research Institute (MDRI) in collaboration with the United Nations Development Programme (UNDP) in September 2020 found that 97 percent of respondents rated the National Steering Committee's pandemic response as either very good (73 percent) or good (24 percent).[39] Unlike in the United States, my Vietnamese friends and interlocutors did not view universal masking or stay-at-home orders as the curtailment of their civil liberties, and they struggled to understand how mask rules could erupt in such conflict and even violence elsewhere.[40] The people I knew and spoke with emphasized the collective good and collaborative efforts needed to curb the spread of the virus, viewing short-term sacrifices as a path to a quicker return to normalcy. Their motivations for masking and adhering to protocols stemmed not solely from self-interest, but from a desire to protect their families and the wider community. As one interlocutor explained in simple yet powerful words, "My actions impact others." Rather than viewing herself as a strictly autonomous individual, she understood her position in society as intrinsically relational, embedded in web of social and ethical obligations. She was acutely aware that failing to comply with regulations could trigger a chain of unfortunate events far beyond her immediate relationships.

Narratives of Vietnam's successful containment of the virus often risk idealizing or even essentializing compliance, however. These accounts tend to reduce complex phenomena to static products of culture, material conditions, or their interaction, fostering a notion of Vietnamese exceptionalism that portrays the country as fundamentally different from "the West." Some scholars attributed Vietnam's widespread adherence to pandemic measures to Confucian influences, describing a Confucian tradition of collectivism as the basis for an ethics of care that prioritized community welfare over individual well-being (Ivic 2020, 345; Vo et al. 2020). This analysis suggests that compliance stemmed more from passive acceptance of fixed cultural norms and predetermined values than from active decision-making in response to evolving circumstances and priorities. Other observers, viewing compliance as coerced and similarly devoid of agency, invoked Orientalist tropes of authoritarian tendencies or strongperson rule, overlooking the possibility of a democratic impulse to safeguard collective life.[41] Both perspectives fall short of adequately capturing the dynamic historical and cultural forces that have shaped Vietnamese late socialist society. A nuanced analysis must consider critical factors, such as changes in political economy, which have led to increased public-private investments in health care and other essential infrastructure. Additionally, Vietnam's historical experience with emergency preparedness manifested as another key factor shaping the country's response. Prior encounters with what I term "disaster socialism" and its associated sonic practices had likely enhanced public receptivity to centralized health directives.

In all instances, compliance, whether compulsory or voluntary, tended to be overstated and framed as absolute. In my experience, a broader spectrum of conduct better reflected the reality on the ground, which evolved over the two-week period of lockdown. Compliance was publicly performed and privately negotiated. Spatialized, it possessed its own temporal rhythm to its ebb and flow. There were workarounds—consider the flexibility embedded in the language of Directive 16. One could comply with measures while flouting them in a single place, action, or moment in time Compliance and noncompliance were thus in constant tension with one another, mutually reinforcing, and often overlapping. Take, for example, permission to leave the house to procure food and other essential goods. Unlike the stricter, Delta-variant lockdown in 2021, which rationed shopping permits for Hanoi households, the April 2020 measure did not specify restrictions on when or how often one could go to the market. In my household, we sometimes visited the marketplace twice daily, once in the morning and again in the early evening, at off-peak hours to avoid the lines that formed outside taped-off shops and stalls to protect vendors. We typically took the long way home, adding an extra block to our route, even though the market was only a two-minute walk away. At times, I would slip out under the guise of a shopping trip to get some fresh air. I made sure my "prop" purchase—a few tomatoes or pieces of tofu—hung visibly to demonstrate compliance if I happened to encounter the ward's "urban order" officer. I was not alone in my non-compliant compliance; I eyed neighbors who passed me on the other side of the street with small, pink plastic bags dangling from their wrists as they, too, took short lockdown walks to secure their basic needs.

As the second week unfolded, I perceived a shift in the rhythmic dynamic between compliance and noncompliance, as the sonic ecology of urban time-space evolved. Erving Goffman's theatrical metaphor of "front stage" and "back stage" ([1956] 1990) offers a useful starting point for examining new patterns of sonic conduct that emerged within the horizon of action and agency, though here I emphasize the fluidity between these increasingly blurred distinctions.[42] Goffman's concept of social life as a performance of norms—observed, heard, and potentially disregarded—aligns with perspectives from feminist scholars of race and gender, who examine how individuals present and navigate their subjectivities across diverse contexts of power and subjection, often subversively (e.g., Butler 1990; hooks 1984).

I extend these ideas of performativity to the pandemic's sonic theater by examining variations in acoustic life across two porous realms: the frontstage, where authorities created and curated official soundscapes while still leaving room for noncompliance, and the backstage, composed of elusive sonic spaces where modulation of frontstage events made obscured sounds progressively audible. This framework offers insights into the performative nature of compliance alongside the potential for what Salomé Voegelin calls "alternative sonic life-worlds" that generate defiant sociality (2021, 33), echoing Attali's ([1985] 2009, 122) observation that noise can resurface as subversion. On the frontstage of public sound—the domain of oratorical performance, in this case, public broadcasts—people within visual or auditory range of one another conduct themselves in a manner that seemingly "maintains and embodies certain standards" (Goffman [1956] 1990, 110), such as adhering to official mandates. In this context, I observed expressions of compliant, care-full listening that demonstrated rational dispositions and moral subjectivities befitting a public health emergency. Offstage, in the alleys away from the public eye and ear, where "the impression fostered by the performance is knowingly contradicted" ([1956] 1990, 114), one found respite from being seen and heard, along with greater latitude to deviate from expected listening norms. Here a potentially subversive sociality unfolded; backstage is where sonic dissent emerged.

Sonic Dissent

While frontstage served as a space for sonic conformity and the performance of public listening, backstage evolved into an acoustic realm of noise and "nonadherence," a concept that medical anthropologist Nicole Charles develops to highlight the ways individuals actively and rationally make choices about their health that may appear to others as careless (2022, 6). "Noise" in this context refers to undesirable sounds and vibrations—perceived auditory disruptions or, in the case of the pandemic, acoustic matter out of place and time. Rather than existing as objective reality, noise relies on the "vagaries of human perception," subjectively measured and relationally defined (Peterson 2021, 99). Beyond its sensory and affective qualities, noise can act as a catalyst for social or political transformation, driving anticipatory, if not outright emancipatory practices (Attali [1985] 2009, 123). Urban noise in particular, because of its potential for subversion and social disturbance, has historically been subject to surveillance and control (Cardoso 2018; Hsieh 2021), especially when produced by racialized bodies (Lippman 2021). Debates about urban soundscapes—how cities should sound and who holds the power to decide or intervene—reflect broader racial, political, and socioeconomic struggles over urban citizenship and rights to belong to the city, including the right to sound and be heard.

There is little sonic policing in Hanoi. Despite existing noise ordinances, which reflect the growing importance of ownership and property rights—such as evening bans on loud music or restrictions on excessively noisy construction—enforcement is lax and observance is rare.[43] However, authorities seem to be less concerned with regulating noise as a matter of public welfare, such as the late night karaoke ban, than with disciplining unauthorized practices that threaten social order through their sounds.[44] Nevertheless, people often find ways to circumvent rules and regulations through both subtle and overt transgression, commonly known in Vietnam as "fence-breaking" (*phá rào*). Fence-breaking is often tolerated at local levels of governance, where a political culture centered on interpersonal relationships (*quan hệ*) and mutual understanding (*thông cảm*) manifest informally between residents and authorities.[45] I have referred to this dynamic elsewhere as "empathetic bureaucracy" to highlight how affectively attuned governance shows surprising flexibility in administrative actions that are responsive to human needs (Schwenkel 2020, 276).[46] Despite the appearance of strict control, this discretionary approach creates spaces—including acoustic ones—for negotiation and accommodation, which shape the contours of everyday life and livelihoods by easing the rigidity of official policy, though not always equitably.

The COVID-19 pandemic revealed similar dynamics. Notwithstanding the initial adherence to policies I observed in Trúc Bạch Ward, there was a perceptible change in my neighbors' offstage activities during the second week of the lockdown. As fear of the deadly virus waned due to its containment, the balance gradually tilted: individuals increasingly prioritized personal and social needs—including economic well-being—over public health security. This nonadherence took form to sound as sonic dissent—hidden from the eye, but unmistakable to the ear.

I use the term "sonic dissent" not to draw a parallel with the acoustic properties of political activism, such as protest songs that challenge government legitimacy (Tausig 2019) or confront white supremacy through a sonic history of black resistance (Brooks 2006).[47] Moving beyond "presumptions about what dissent is *supposed* to sound like," I shift my aural attention to those sonic expressions that enact more subtle forms of audible nonconformity (Tausig 2019, 26, emphasis in the original). As a manifestation of agency that neither resists nor rejects state policy, sonic dissent in pandemic Hanoi could be heard as an acoustic deviation from regulations, a fence-breaking-through-noise-making maneuver that straddled the line between compliance and nonadherence. Through offstage actions, sonic dissenters boldly departed from COVID-19 directives, creating an alternative soundscape of sociality that echoed a longing for a post-pandemic normalcy. While this noise-based defiance subverted established power hierarchies (Porath 2019), it did so by challenging the rules rather than the ruling authority.

Sonic Sociality: The Party

During the second week of the lockdown, the frontstage of pandemic cultural norms remained sonically unchanged. Across the street from my building, the soothing sounds of fountains aerating the lake to improve water quality punctuated the silences between ritualized messages of authority that marked the start and end of the day. Backstage, however, behind the building, in a narrow alley, a different acoustic environment emerged. There, increased noise levels hinted at the possibility of a return to routine daily life, to predictability unmarked by crisis. This collective sounding created the conditions for sonic sociality, drawing some of my neighbors out of isolation to re-establish communication. Unlike the one-way transmission of the loudspeaker on the frontstage, this offstage space offered more interactive auditory encounters, allowing listeners to both speak and hear.

It began with a noontime *nhậu*, or drinking party that typically accompanies social eating. A back alley wound its way behind my flat, flanked by narrow houses, some with ground floor businesses, culminating at a scrap-metal collection point and ambulance transport facility. On special occasions, such as national holidays, this spacious lot transformed into a lunchtime *bia hơi*, or makeshift draft beer garden, complete with low plastic stools and tables, a keg, and several hot dishes prepared by my neighbor, Giang (figure 19). When we first met, Giang, like many women vendors who appropriated common spaces for their livelihoods

FIGURE 19. Alleyway acoustics: sound echoed in the enclosed space between my back wall (3 o'clock, with the rolled down door), the metal recycling site (12 o'clock, where a scrap collector stands with her bike), the vacant lot repurposed as a periodic *bia hơi* (11 o'clock, behind the green door), and my neighbor's house (9 o'clock), which appropriates the commons for her lunch business, merging public and private space, 2022. Photo by the author.

(see Schwenkel 2022b), ran a *bún chả* (grilled pork and rice noodles) stand from her home. However, during the pandemic, to make ends meet, she turned her small kitchen into a fruit and vegetable shop, delivering fresh produce to neighboring households.[48]

Giang's pre-pandemic *bia hơi* lunch gatherings were boisterous affairs. Shouts bounced off the contiguous walls, amplifying as they reached my seventh-story office windows. As Giang's guests became inebriated, the revelry would grow louder and stretch on for hours. To maintain my work focus amid the rising decibels—and to sleep through the midnight racket of scrap iron hurled into trucks—I resorted to purchasing a white-noise machine. The March outbreak in Trúc Bạch Ward put a stop to these rowdy gatherings, and I enjoyed an extended period of relative quiet in my workspace through the first week of lockdown.

That changed without warning. One quiet afternoon, an unexpected echo broke the silence of isolation. I heard, loud and clear, the lively utterance of what had become unutterable in our state of emergency: a toast—"Một, hai, ba, dô!" (one, two three, bottoms up!). If, as Astrid von Busekist (2001, 109) argues, "modes of exercising power . . . are materialized in the unutterable," the open utterance of an expression associated with forbidden actions (social drinking during lockdown) directly contested that power by audibly sounding dissent. To my surprise, the noise swelled, and the party grew rowdier, a cacophony of togetherness in spaces meant for solitude. Excited voices filled the air as men and women raised their volume, competing to be heard. Their sonic sociality expressed itself through high-spirited revelry as they challenged one another to *"một trăm phần trăm"* ("one hundred percent"), urging each other to empty their glasses in one go.

Intrigued by the brazen defiance of lockdown, I opened the window and peered down, but the source of the clamor remained hidden. Without a direct line of sight, I strained to listen. The familiar clinking of heavy glasses punctuated what sounded like a sizeable gathering, far exceeding the two-person restriction in place at the time. Given the crisis situation we were in, I anticipated imminent intervention by authorities. Yet no one arrived, and the commotion continued unabated. Could the civil servants at the nearby People's Committee, just a hundred feet away, be oblivious to the sounds emanating from this shared social space? Or perhaps the scale and brief nature of this nonconformity did not warrant sufficient concern? Could the sonic camaraderie unfolding backstage, within earshot of the police stationed across the canal, truly escape their notice? Or was the state simply not listening? Offstage violations of pandemic mandates, concealed within the liminal spaces of alleys behind closed gates, appeared to be tacitly tolerated in my neighborhood, evading government scrutiny despite the high risks.

Sounding Normalcy: Balcony Renovation

Sonic dissent did not end with the ephemeral afternoon party. Soon after, a neighbor with an adjoining property reintroduced everyday sounds into the auditory experience of the city—this time through an extensive renovation of his home. Construction noise in Hanoi, an insufferable outcome of rapid urban development, contributes to Hanoi's dangerous levels of air pollution, the second highest in Southeast Asia. Daily red-coded alerts from the Vietnamese social media platform Zalo, warning of unhealthful or "hazardous" levels of pollutants (*"mức có hại cho sức khỏe"*), including high concentrations of fine particulate matter (PM 2.5), have become a grim reality of urban life.[49] The lockdown afforded the urban environment a temporary reprieve, even as it further exacerbated the precarity faced by nonessential workers, such as ride-hailing drivers (see pandemic figure 4). Air quality improved considerably as construction and transportation ground to a halt, leading to a noticeable reduction in emissions as the toxic dust cleared. Noise pollution levels followed accordingly, basking Hanoi in a rare ambient stillness that accompanied an atmosphere of more breathable air.

Recuperation would be short-lived. Construction work on my neighbor's house commenced in the early morning during the second week of lockdown, marking a sonic turning point in the gradual shift to a new urban normal—and the end

AUDIO CLIP 4. Sounding dissent: construction noise from home renovation signals noncompliance during the second week of lockdown amid a loosening of restrictions. Source: author.

To hear this audio clip, scan the QR code with your mobile device or visit DOI: https://doi.org/10.1525/luminos.249.audio4

00:00:02 —— 00:00:29

Audio clip 4

to urban quietude. The jarring jackhammering, piercing steel-cutting, relentless banging, and persistent pounding would continue for weeks as workers enclosed two upper-floor balconies and added an extra story to the private residence. This cacophony of construction took place offstage, or rather, above-stage, over the city—at eye and ear level with my apartment (audio clip 4). From the ground, the remodeling project was as hidden as the alley party—easy to hear but difficult to see, unless one looked up high enough, at just the right angle, through the trees. Its soundscape, however, was undeniable. Reaching every ear, it dramatically altered the acoustics of the locked-down city, even rivaling the amplifications of the loa phường.

The labor-intensive building industry is inherently collaborative, with sound-making a collective activity—no one builds or populates a construction site alone. In this instance, the scale of sonic dissent raised questions about the planned logistics of nonadherence, contrasting with the more spontaneous sonic sociality of the drinking party: How did a group of male and female migrant workers, who form the backbone of the industry, manage to reach Hanoi despite the tightly enforced travel ban?[50] How was it possible for them to live and work together on-site, given the explicit prohibition of nonessential operations, including construction activities, and the strict distancing measures imposed by Directive 16? And how did my neighbor orchestrate the nighttime delivery of building materials, during which I could hear steel piles being hastily tossed from a truck onto the street, causing the ground to shake before they were hoisted into place?

These two sound events, occurring across different social classes—the former in a precarious household and the latter in a more affluent one—call for a more nuanced approach to the narrative of Vietnam's successful virus containment through universal compliance with pandemic mandates.[51] While public displays of social cohesion and shared values of care were evident on the frontstage, an ethos of the common good collided at times with the pursuit of self-interest offstage, materializing through sonic counter-practices. The fact that this pandemic "fence-breaking" was heard rather than seen underscores the critical role of the ethnographic ear and participant listening in fieldwork—or, more aptly, soundwork. During the 2020 lockdown, attentive backstage listening revealed that people's commitments to social responsibility fluctuated spatially and temporally, particularly in the offstage spheres of daily life, beyond public hearing. Noise as dissent signaled a rejection of pandemic time, conveying a sense of urgency to accelerate the arrival of a post-COVID city. By creating a "sonic possible world" (Voegelin 2021, 13), this dissent envisioned an alternative reality free from the threat of a public health emergency, a possibility expressed and made manifest through sound.

Despite their audaciousness, these examples of backstage, sonic fence-breaking were not altogether exceptional. During the second week of the stay-at-home order, I came to see the ubiquitous rolling security door—a common residential and commercial feature in Hanoi, which remained mostly shut during the first week of social isolation—as a visual and auditory symbol of the incremental return to normalcy through widening forms of noncompliance. The creaking ascent of steel doors cautiously opening—initially just a few feet, but then gradually higher—hinted at a waning fear of infection among residents and business owners eager to resume life and recoup lost profits. It also suggested a small but steady pushback against restrictive state policies. *Phở cuốn* (phở rolls) shops in my neighborhood quietly resumed takeout and delivery, subtly signaling their reopening by partially raising their rolling doors. This shift reintroduced the familiar sounds of motorbikes to the urban soundscape while burdening the environment with large volumes of single-use plastic waste.[52]

I witnessed this accumulating waste firsthand during evening visits to the garbage cart parked across from the market. These occasional trips afforded my spouse and me brief walks through the desolate neighborhood, hours after the cart had passed our flat. Its loud bell, one of the few signs of routine infrastructure operations, served as both an alert to its presence and a pretext for our nighttime ventures. We were not alone in these brisk strolls through silent streets under the cover of darkness. Though we maintained distance, other couples also visited the rubbish cart and then wandered into the shadows beyond in acts of noncompliant compliance (video 2). The previously unfamiliar sight of couples walking or biking in the evening highlighted how the pandemic altered the gendered use and practice of public space—a topic I explore in act 3 with social distancing protocols.

On April 15, the government extended its two-week shelter-in-place order for an additional week in high-risk areas, impacting several provinces and cities, including parts of Hanoi and specifically my ward. Sensing growing leniency from authorities, residents continued to bend the rules while remaining mindful of public health obligations, confident the city would avoid another mass outbreak. With no new infections reported anywhere in the country for a week, and a total case count holding steady at 268, the nationwide lockdown was brought to a close on April 22. Slowly, the city came back to life. Domestic flights on major routes resumed a few days later, and public transportation restarted in May. Businesses were officially permitted to return to normal operations, though restrictions remained for some high-risk venues, like bars and karaoke lounges. The prohibition on itinerant street vending persisted, revealing how emergency powers were

VIDEO 2. Evening jaunt through deserted streets at the start of the second week of the lockdown in Trúc Bạch Ward, 2020. Video by the author.

To watch this video, scan the QR code with your mobile device or visit
DOI: https://doi.org/10.1525/luminos.249.video2

used to reinforce exceptional regulations that risked criminalizing the livelihoods of vulnerable segments of the population (Young 2021, 989).

For the next ninety-nine days, Vietnam would record no community transmission of COVID-19 until July 25, an impressive feat at a time when the United States had reached four million infections and well over one hundred thousand deaths.[53] During these three months, which I detail in the next act, the country entered a period of compulsory distancing that altered how people used, navigated, and interacted with sound in public spaces. While outdoor activity was no longer perceived as an infringement—and was even encouraged to maintain health and fitness—the state's acoustics of power remained in effect. After all, the virus could reappear at any moment. This "new normal" (*bình thường mới*), shaped by spatial governance, brought changes to the sensory qualities of the city, which I experienced through *soundwalking* as embodied research practice. Walking—particularly in the form of "sensory walkabouts" attuned to the sonic environment (Low and Abdullah 2021)—offered a perception-driven approach to reconnect with the city's revitalizing soundscape. This method required a shift from isolated self-care to mutual care in co-created spaces, as people transitioned from co-present listening in separate homes to co-present walking outdoors—from listening with to walking alongside, while striving to maintain sufficient interpersonal distance.

Nonessential Workers

The Grab Driver (Anh Grab)

The first nationwide lockdown in April 2020 successfully halted community transmission of the coronavirus, but it also exposed the social and economic vulnerabilities of a precarious population: nonessential workers in the gig economy. As independent contractors typically lacked job security and the social protections extended to full-time employees, gig workers were disproportionately affected by the strict government measures implemented during the pandemic. There were exceptions, of course. While some online platform workers, like those in food delivery and e-commerce, were able to capitalize on new business opportunities, others faced complete shutdowns. This was especially true for ride-hailing drivers in Hanoi, who experienced a complete loss of income during the extended stay-at-home order.

Disruptions to urban mobility radically transformed the city's sonic atmosphere. As discussed in act 2, state emergency measures in response to COVID-19 reshaped Hanoi's soundscape, with streets emptied of vehicles and social activity. Noise pollution waned as the familiar sounds of squeaking brakes, relentless honking, revving engines, and sputtering mufflers—hallmarks of Hanoi's chronic traffic congestion—all but disappeared. For some residents, this cacophony had been a source of urban disturbance, but for others, like Grab drivers, these very sounds symbolized livelihood opportunities. Under Directive 16, the abrupt silence in the nonessential transport sector became a stark acoustic marker of economic devastation, as app-based services like Grab, the Southeast Asian giant of the ride-hailing industry (Panimbang 2021), came to a standstill.

The explosion of ride-hailing apps in Hanoi over the past decade has had a tremendous impact on how people move around the city (Turner and Ngô 2019).

GrabTaxi entered the market in 2014 and rose to dominate the e-hailing sector following Uber's exit from Vietnam in 2018, rapidly supplanting the traditional taxi industry.[1] Overnight, Hanoi's ubiquitous Mai Linh and Thành Công cabs seemed to vanish from the streets as drivers, lured by the appeal of "flexible time" (*thời gian linh động*) and "reliable income" (*thu nhập tin cậy*), shifted their loyalties to app-based transportation services. This shift occurred despite constant monitoring through the very data-extractive technologies that afforded drivers the illusion of "freedom" (*tự do*) to self-manage and "take control of their lives" (*làm chủ cuộc sống của mình*).[2]

The pandemic amplified the vulnerability of this predominantly male workforce (hence the colloquial term "*anh* Grab," or brother Grab driver), who were framed as "partners" (*đối tác*) in a presumably mutually beneficial, horizontal relationship with companies, rather than as employees occupying a subordinate position within a power hierarchy. This arrangement, emblematic of platform capitalism, externalized risks and responsibilities onto the drivers, leaving them exposed to the uncertainties of a digital sharing economy (Vallas and Schor 2020). The precarity of many ride-hailing drivers was compounded by their marginalized status as rural migrants who lacked permanent residency (*hộ khẩu*) in the capital city and, as a result, were ineligible for government relief payments meant for households adversely impacted by the pandemic.[3] Moreover, the firms themselves offered little financial or other assistance to their driver-partners. Lacking a social safety net and relying solely on familial support, nonessential gig workers for taxi platforms faced the dual threat of losing both their income and their primary capital investment: their cars. For taxi drivers, the silence that swept across Hanoi's streets during the lockdown was not merely an acoustic event but the sound of their suspended livelihoods.

Sơn's life circumstances and story echoed those of many other Grab drivers I spoke with during the period of social distancing, after restrictions on nonessential travel were lifted in May.[4] Sơn's path to financial autonomy was contingent upon automobility and access to credit. Several years before the pandemic, he had left his family in Bắc Ninh Province to work as a taxi driver in Hanoi, where he shared a dormitory room with three other migrant drivers, each paying one million VND (US$43) for rent and utilities. From his monthly earnings of approximately ten million VND (US$430) as a Mai Linh taxi driver-employee, Sơn judiciously saved until he had enough for a down payment on a bank loan to purchase a new Toyota sedan—his first privately-owned car. His goal, he told me, was to become a "freelance worker" (*lao động tự do*) for Grab.

Sơn saw this transition to debt financing as enabling social mobility, and his income would double to an average of twenty million VND (US$850) per month after making the calculated move to driver-partner. He liked the idea of controlling his work hours and schedule; before the pandemic, he usually shut off the app during rush hour, which was less profitable given the time spent in Hanoi's notoriously gridlocked traffic. However, he disapproved of Grab's high commission taken from his earnings—20 percent plus taxes at the time—and was concerned about how pandemic fare hikes, surcharges, and rising gas prices would affect his net income. What initially sounded like an endorsement of platform capitalism's persuasive promise of liberation (Srnicek 2017, 44) swiftly turned into recognition of its propensity for exploitation.

Sơn was a diligent driver, motivated to pay off his loan as quickly as possible, even if it meant working extra hours. His eye was on saving money for his children's education, for both his daughter and son. The purchase of a private car, once unattainable for most Vietnamese families a decade ago, became possible due to reduced tax rates and fees for domestically assembled vehicles, coupled with greater loan accessibility, which made car ownership more affordable in Vietnam (Hansen 2017).[5] While some Grab drivers I spoke with paid cash for their vehicles, often borrowing from relatives, Sơn financed his car with a four-year bank loan at a remarkably low interest rate of 1.6 percent. His monthly payment was eight million VND (US$340), higher than the more typical five million VND (US$215) reported by other drivers, accounting for 25 to 30 percent of their net pay. "I agreed to a higher payment over fewer years because I don't like debt," he told me one August morning. This meant longer driving hours, including weekends, instead of returning home to see his family, which he managed to do just once a month. But the pandemic upended his sound financial planning. With just six months left until he was free of debt, Sơn found himself without income and unable to make his next car payment.

Like many Grab driver-partners, Sơn was meticulous about his car and about delivering excellent service through small acts of customer care to ensure a smooth ride. He beamed with pride when I complimented his sleek, well-maintained sedan and the comfort of its black leather seats. On the front dashboard, Sơn had placed a faux jade Buddha talisman, a cheerful Di Lặc with a rounded belly, believed to offer protection against accidents and illness, in addition to bringing prosperity. In the middle console, he had arranged a modern-day amulet—a bottle of hand sanitizer to ensure his passengers' well-being. Although we both wore face masks, muffling our conversation, there was no plexiglass divider between us, as became common in places like Hồ Chí Minh City.

The pandemic altered the social dynamics between rider and driver, muting the usual ambient noise of our shared space. Typically, after my phone rang signaling the Grab driver's arrival, I would rush outside to the waiting car and confirm the license plate before jumping into the front seat. Now, however, I sat in the back to maintain some semblance of social distance, at times in awkward silence, partly because many Vietnamese associated COVID-19 with foreigners. I knew my presence could add to drivers' sense of occupational risk, despite Hanoi being largely virus-free at the time. Sơn himself had speculated that the recent Đà

Nẵng outbreak, which ended a 100-day streak without community transmission, had been caused by illegal Chinese migrants who had bribed corrupt Vietnamese border guards to gain entry into the country.[6] Sơn's suspicion reflected the general unease that pervaded the social atmosphere, stifling the once-lively exchanges that had animated taxi journeys.

Yet in a moment of connection, Sơn and I chatted amicably about the April lockdown during the long ride to Thanh Xuân District in southern Hanoi. He had returned to his home village to quarantine with his wife and children since nonessential transportation had been banned and he was not permitted to work. All his roommates, in fact, had left the city behind. Sơn scoffed when I asked about the government's relief program. No driver I spoke with had qualified for state support given their status as independent contractors, and Sơn was no exception. A few of his coworkers might have been eligible, he surmised, but no one he knew personally. Grab management had offered a loan to sustain him during this time of hardship, but he thought it foolish to accept a loan to pay another loan: "I already have bank debt!" (*Nợ ngân hàng rồi!*). He also refused the in-kind support that Grab supplied to combat food insecurity: five kilograms of rice and a box of instant noodles, which he considered a pittance. "I needed meat and vegetables, not rice," which his family grew, he bemoaned.

Sơn continued to struggle financially after returning to Hanoi in May. Although domestic travel restrictions had been lifted, the number of riders had declined considerably. There were few expatriates and no international tourists, Grab's major clientele at the time (up to two-thirds of e-hailers, drivers informed me). With schools closed and employees still working remotely, Hanoians rarely ventured far from home for nonessential reasons. Sơn estimated a 50 percent decline in ridership. The streets, though no longer completely silent, lacked their pre-pandemic intensity; the subdued urban rhythm became an acoustic reminder of diminished income opportunities that defined the new normal. Drivers reported having to work longer—up to twelve hours—to make the same pay, averaging twenty to thirty passengers per day. But even then, making ends meet proved difficult. "People don't go out much these days," Sơn lamented, "and if they do, they tend to avoid taxis."

Act 3

———

Social Distancing

Mutual Care

Don't spread false information!
Maintain your social distance, and wear a mask any time you go out . . .
Don't stigmatize, be civilized!
Remain calm, we will get through it all.
All the nation, gather as one,
Vietnam leaves no one behind!

—"HEY VIETNAM! LET'S FIGHT COVID!," MINISTRY OF HEALTH
PUBLIC SERVICE VIDEO TURNED MORNING WORKOUT ANTHEM

If staying at home expressed love of country ("ở nhà là yêu nước"), then maintaining distance ("giữ khoảng cách") in public showed care for fellow citizens. As both an ethical posture and a policy of containment, social distancing regulated the intersubjective interactions among human and other-than-human bodies, and in so doing restricted the exchange of sensations that occurs in shared public space.[1] Observing the mandated physical distance of two meters, or six feet, demanded a mode of "border thinking" as a gesture of solidarity involving mutually embodied experiences of vulnerability (Icaza 2017, 33), a gesture in which management of self in relation to others assured protection for the social good. To navigate "safe" space was to calibrate the risks posed by varying degrees of sensory distance and proximity. The potential of a cough, sneeze, or even loud talking to release viral droplets into the air became a *sonic* matter of epidemiological concern, requiring people to attune themselves to distinctive bodily sounds as "auditory prompts" or warnings of possible illness to avoid (Brown et al. 2021, 280).

The end of the lockdown marked a shift in pandemic governance and its sonic dynamics, moving away from emergency interventions, or states of exception involving social isolation, to a focus on anticipating and controlling future outbreaks. This occurred as the country transitioned into a phase of enforced spatial distancing, designed to minimize close physical contact and social interaction in

98

both open and closed environments.[2] These public health measures reshaped the urban atmosphere, altering the material, spatial, and sonic characteristics of the city, and, in turn, transformed people's perceptions and experiences of place. For example, the recommended two-meter separation was frequently indicated on the ground, using visual cues such as red squares, white lines, or painted yellow footprints, and was reinforced by the amplified instructions of bullhorns and other sound reproduction technologies. Restrictions on gatherings remained in place, with signs posted by authorities indicating the maximum number of participants allowed based on location, type of activity, and the potential for respiratory aerosol transmission. Remote work and home schooling continued, keeping ambient noise and traffic to a minimum, while many small businesses remained shuttered until the owners returned to the city. Those establishments that did reopen created improvised barriers to demarcate the boundaries between inside and outside, familiar and strange, safety and danger. Efforts to remap and control access to space sought to regenerate livelihoods by mitigating the risks of viral contamination through market transactions.

For the most part, people took seriously the need to physically distance from one another, showing how care work and cohabitation were deeply intertwined and mutually constituted in the rhythms of daily life. Lingering uncertainty about contracting and unintentionally transmitting the virus to others—be it family members, endangering elderly relatives, or the wider public—impacted how people navigated the city spatially and temporally in post-lockdown society. It also shaped the dynamic production of sound in and through social interactions, as I demonstrate in this chapter. A coronavirus love poem (*thơ tình thời Covid*) posted online, which adapts a stanza from Chế Lan Viên's 1937 poem "Đêm tàn" (Brutal Night) to fit the context of the epidemic, captures the fear of cross-contamination between young lovers in close proximity who are afraid to speak, much less breathe:

> We look at each other without saying a word
> Afraid of opening our mouths lest the virus spread between us
> We hold our breath in silence
> Twin souls immersed in a sea of sadness[3]

The possibility of targeted media coverage heightened public anxieties about acquiring and spreading the virus, especially given the state's exhaustive documentation and contact tracing efforts for every instance of community transmission. Being in the wrong place at the wrong time could unexpectedly land someone in centralized quarantine for two weeks, even if asymptomatic, an inconvenient disruption at best.[4] Worse, it could result in being featured (though not openly identified) in news reports, ostensibly to protect wider society, as happened with Patient 17 (see pandemic figure 1). This likelihood was particularly high if one's actions were deemed reckless, drawing unwanted scrutiny and stigma, culminating in the public withdrawal of compassion.[5]

The auditory reminders to increase physical distance while reducing social contact were persistent, in case individuals might overlook such precautionary measures and violate the tacit "pandemic contract" to exhibit care and solidarity by respecting those agreed-upon boundaries of separation. Sonic devices—including whistles, megaphones, and loudspeakers—were deployed to shape and regulate social engagements and mobility patterns in urban space. These residual technologies, central to sonic socialism, intertwined care and control, revealing how media infrastructures simultaneously fostered both public health surveillance and collective solidarity. "Do not gather in groups!" "Suspend exercise and sports in public spaces!" "Maintain a minimum distance of two meters!" "Avoid crowded places!" "People with suspected symptoms should not travel!" These were among the audio messages amplified around me as I ventured back outside.

The lifting of lockdown restrictions saw a surge in urban mobility as people left the confines of their private homes for the communal spaces of the urban outdoors. Health-conscious forms of mobility, such as bike riding, which aligned with distancing measures, gained immense popularity among the middle classes in my neighborhood. The spike in demand quickly depleted local bike shop inventories.[6] Other activities disregarded directives. Women resumed their morning fitness routines, despite signs prohibiting group activities, and men gathered for team sports, like *đá cầu* or foot badminton, occasionally with co-ed participation.[7] The collective enthusiasm for fresh air, green spaces, and social reconnection was palpable as my neighbors, often accompanied by their dressed-up canine companions, eagerly took to walking, relishing the renewed potential for co-presence it provided (figure 20).

On April 22, 2020, I set out on a morning walk around Trúc Bạch Lake. Because of my quarantine (act 1) and the ensuing nationwide lockdown (act 2), more than a month had passed since I last traversed the entirety of the mostly shaded, thirty-minute route (figure 21). The cool spring air felt crisp in the early dawn hours, prompting me to zip up my jacket as I stepped onto the street. I breathed deeply before donning my disposable mask. Surveying my surroundings, I paused to absorb the morning COVID-19 announcements being broadcasted through the ward loudspeaker adjacent to my building. From across the lake, the echoes of a rival PA system with similar public health messaging reached my ears. I observed a couple of older passersby, who kept their distance from one another and from me, a foreigner they regarded with suspicion, possibly carrying the airborne disease. The street was otherwise still, devoid of vehicles with the exception of a few bicycles. I turned right, took a few strides to join the pedestrian flow, and began what would become my ambulant research practice of "walking with," an embodied method of sensory ethnography—to thinking, feeling, listening, and attuning to bodies, surfaces, sounds, and movement through urban space (Springgay and

FIGURE 20. Human-nonhuman animal encounters: the family dog accompanies their owner to morning cardio sessions at Trúc Bạch Lake, sporting a different outfit each day during spatial distancing, 2020. Photo by the author.

FIGURE 21. Trúc Bạch Lake with walking path, which follows the contours of the water's edge under the cover of fragrant orchid, flame, and banyan trees. Red dot indicates the author's starting point and residence. Total distance is approximately 1.6 miles. Source: Google Earth.

FIGURE 22. Release from lockdown: eager fitness enthusiasts bypass barricades to access park amenities, despite posted notices banning such activities, April 22, 2020. Photo by the author.

Truman 2019, 142). Pandemic walking—especially soundwalks—manifested as an intersubjective experience of "with-ness" while respecting spatial distance. As a relational practice, it afforded opportunities to co-sense alongside fellow bodies, both human and nonhuman, while navigating routine activities and engagements with urban life (Low 2015).

Walking in a pandemic both affirms and disrupts relations of power and attempts at social order through sensory and bodily discipline. While making my way around the lake that first morning, I passed the metal barricades erected by the police at the start of the lockdown to control pedestrian movement. Still firmly in place, they stood as a reminder of the subversive potential of walking (Solnit 2001, 285), or strolling while defying the rules of the street.[8] Affixed to these barricades were red vinyl banners with yellow block lettering, purposefully hung by the Ward People's Committee, prohibiting the gathering of groups and the use of lakefront amenities, like the exercise machines that are popular with my neighbors.

On the open stretch of Thanh Niên Street, a pair of traffic officers patrolled the pavement, ensuring that pedestrians kept moving and maintained spatial distance, while keeping an eye and ear out for the cries of street vendors, who were slow to return to the city. Away from the main thoroughfare, however, people were more prone to ignore the signs, despite the posted fines and acoustic reminders of proper conduct outdoors (figure 22).[9] Gradually, the dynamic between compliance and nonadherence—or accommodation and refusal—shifted as aspirations to self-actualization clashed with health protection policies. Within a matter of days, new forms of sonic dissent emerged through the reverberating rhythms of early morning workout music as people, especially women, made acoustic claims to the city. These spaces of sonic belonging marked the onset of a deceptively "new normal" or bubble of safety that resounded with hopeful affects across the city.

VIỆT NAM ƠI! (HEY VIETNAM!)

One month after the release of "Ghen Cô Vy," the Ministry of Health collaborated with musician Minh Beta to produce "Việt Nam Ơi! Đánh Bay Covid" ("Hey Vietnam! Let's Fight COVID"), a song and music video that would become central to the sonic experience of spatial distancing.[10] Similar to the way "Ghen Cô Vy" reimagined the hit song "Ghen," Minh Beta's pop-sonic, public service announcement was a pandemic-inspired rendition of his 2019 smash hit, "Việt Nam Ơi!" A patriotic anthem celebrating national unity, "Việt Nam Ơi!"—often played at soccer matches and other mass events to stir the enthusiasm of crowds—was repurposed during the pandemic to evoke shared emotions and foster mutual solidarity.[11] Later on, the song found its way into other contexts, notably the morning exercise groups that congregated along the lake in my neighborhood. Eager to shake off the isolation and inactivity of the past weeks, these groups transformed the song's intention from a mere call to action into a purposeful act of reconnection.

In the preface, I argued that "Ghen Cô Vy" formed part of the state's sonic toolkit in the early days of pandemic preparedness. As an audiovisual response to a national health emergency, "Ghen Cô Vy" aimed to encourage hygiene practices grounded in biomedical rationality, while strengthening state legitimacy. Emphasizing interdependency through "musicking" (Small 1998), it created an uplifting shared experience of embodied learning within an imagined sonic public of attuned listeners—and dancing handwashers. By featuring sing-and-dance-along cartoon characters, "Ghen Cô Vy" promoted social cohesion and a strong commitment to collective care as a strategy to curb the spread of COVID-19. This approach cleverly packaged public health imperatives as entertainment, subtly masking the state's paternalism by strategically fusing regulatory measures with popular culture.

"Việt Nam Ơi! Đánh Bay Covid" similarly motivated a public of engaged listeners through sound while obscuring the presence of the state. Like "Ghen Cô Vy," it drew on the affective power of song and dance to activate relations of care among the population and disseminate vital information about the virus while sidestepping the role of power and politics. It, too, boasted a catchy melody and accessible choreography, which gave rise to spin-off videos across a wide spectrum of society, from youth clubs and dance troupes to army conscripts and health care workers, whose synchronized movements enacted a symbolic performance of national harmony. Indeed, one of the primary objectives of "Việt Nam Ơi! Đánh Bay Covid" was to celebrate unity in diversity by spotlighting different social groups around the country who aligned themselves with the state's vision of pandemic optimism, including farmers, factory workers, ethnic minorities, doctors, and youth volunteers. Together, they projected an image of a resilient, homogenized front, devoid of class and ethnic distinctions, allied against an external threat to be struck down and controlled through coordinated displays of competence and care. This celebration of seamless unity, however, masked the disproportionate impact of COVID-19 on low-income and ethnic minority populations, while centering "Kinh-ness" as the cultural standard in pandemic narratives.

Released in April 2020 during the nationwide lockdown, the official video diverges from "Ghen Cô Vy's" focus on proper hygiene practices, like handwashing, in *domestic* spaces of care. Instead, it emphasizes the monitoring of bodily discipline and social conduct that violate public health and safety measures in the *public* sphere. "Việt Nam Ơi! Đánh Bay Covid" thus served as a crucial audiovisual tool for shaping an ethical and relational "pandemic-mandated personhood" (Halstead 2020). Its emphasis extended beyond autonomous self-management to instead highlight the wider social context of individual actions and their repercussions across society. The video features a cast of live-action frontline superheroes, each endowed with a unique power, who intervene to modify behaviors deemed disruptive or counterproductive to the state's emergency response, and which expose the population to heightened risk (figure 23). The infractions targeted for correction are depicted through gendered cultural practices: women exchange folk remedies about garlic and magical amulets to circumvent medical authorities, while men partake in daytime social revelry, consuming beer in close physical proximity.[12]

For example, Lady Realness, clad head to toe in black leather, wields a megaphone to combat popular (non-scientific) beliefs and fake "sidewalk news" (*tin vỉa hè*) that compete for medical credibility and undermine scientific approaches to hygiene. Meanwhile, Mr. Master Mind, a caped crusader fighting against ignorance and self-interest, apprehends a female traveler attempting to evade mandatory quarantine, a reference to the case of Patient 17. Together, the superheroes work to save humanity by bringing order to chaos; they curb hoarding, establish queues, calm panic, disperse crowds, and disinfect the city streets to the tune of "Cùng đoàn kết đánh bay corona" ("Unite to fight corona"). Their efforts help restore civilization by upholding virtuous practices that enhance social productivity and cultivate morally compliant citizens, whose conduct exemplifies the country's "hygienic modernity" (Rogaski 2004). At the same time, the presence of these superheroes—a common trope in pandemic comics—embodies the valor and "health-preserving powers" of real-life frontline workers, even as their selfless

FIGURE 23. Prepared to confront the virus: superheroes monitor pandemic conduct in public spaces to ensure compliance through song. Still from "Việt Nam Ơi! Đánh Bay Covid" with sign language interpretation, 2020.

protection of public well-being is commodified in this representation (Saji, Venkatesan, and Callender 2021, 151).

The music video "Việt Nam Ơi! Đánh Bay Covid" serves as an example of how multimedia infrastructures were leveraged for the sonic governance of safety and care, reminding viewers that "fighting" the virus would require the full participation of rational, hygienic citizens who prioritized collective over individual interests. The emphasis on social mobilization, as the operative logic of disaster socialism, demonstrated that expectations of self-responsibilization were entwined with state protections, rather than disaggregated as neoliberal discourse might suggest. Actions of mutual care and cooperation would form the cornerstone of citizen-state solidarity, instead of their opposition (Trnka 2020, 268). This sonic strategy was in line with the historical use of song as a rallying call to elevate socio-sensory awareness, combat backwardness, and galvanize the population around a common cause or enemy, though here it added a visual dimension to the audio experience. In practice, however, "Hey Vietnam!" did not always elicit compliance; instead, it sometimes prompted individuals to undermine it.

With its strong patriotic sentiments (more pronounced than its musical precursor, "Ghen Cô Vy"), "Việt Nam Ơi! Đánh Bay Covid" deployed an upbeat, infectious soundtrack to promote ethical conduct befitting an enlightened population invested in the public health of the nation. However, the video also hinted at the potential for acts of ignorance or outright refusal to jeopardize the collective struggle, necessitating the intervention of superheroes—as embodiments of state power—to rectify these transgressions. As the song circulated and was adopted in different sonic contexts, it took on a new acoustic life beyond its intended purpose.

"Hey Vietnam!" permeated my daily life during the period of spatial distancing. In many ways, it became emblematic of the sonic identity of the pandemic due to its association with the country's management of COVID-19. The song was acoustically omnipresent. I would hear its melody daily, often from multiple sources simultaneously, and the tune would echo in my mind at other times. Some mornings, the song would commence the 6 a.m. loudspeaker announcements, while on other occasions, it would bring them to a close. The instantly recognizable sound bite, "Việt Nam ơi! Cùng đoàn kết đánh bay corona!" was broadcast by the mobile sound truck making its rounds in my neighborhood. Storefronts and cafés also played the song, allowing its melody to resonate through the streets.

But predominantly and with remarkable consistency, I listened to the song each morning during my daily walk around the perimeter of Trúc Bạch Lake. From the break of dawn, fitness groups congregating in the open spaces along the lake's northern shore incorporated "Việt Nam Ơi! Đánh Bay Covid" into their morning workout routines, turning up the volume on their portable amplifiers to boost energy levels, even as such gatherings violated the clearly posted guidelines. Ironically, despite its emphasis on following the rules, the music was bound up with fence-breaking practices when repurposed for alternative acoustic environments. This supports my argument that the pandemic transcended a mere public health crisis managed through institutional responses and expert interventions. It also emerged as a pivotal sonic event, given the role of sound in managing interactions between humans and nonhumans, including pathogens. This sonic mobilization enabled bodies in motion to actively reshape the pandemic's dominant sensory order in public spaces, as sonic agents appropriated state-produced sounds, adapting them to their own acoustic sensibilities and embodied practices of re-sounding the city.[13]

PACING THE PANDEMIC: SOUNDWALKING

Where does it start? . . . It starts with a step and then another step and then another that add up like taps on a drum to a rhythm, the rhythm of walking.
—REBECCA SOLNIT, *WANDERLUST: A HISTORY OF WALKING*

Walking—the simple, monotonous act of placing one foot before the other to prevent falling—turns out not to be so simple if you're black.
—GARNETTE CADOGAN, *WALKING WHILE BLACK*

Sonic experiences of the pandemic reached a turning point on April 22 when Vietnam wound down its prolonged nationwide lockdown with no recorded fatalities and only 268 confirmed cases of infection. That morning, stepping outside was an act of sensory reawakening. I felt a surge of excitement to reengage the world sensorially, to feel the surface of the earth beneath my feet, to breathe in the air and acclimate my senses to ambient sounds and the presence of other humans and nonhumans—including cutout ones—moving toward me, alongside me, past me, or me past them (figure 24). This cathartic moment of collective being and sensing—of navigating a new normal of distanced sociality as an ethics of care for life—would evolve into my daily ethnographic routine: immersive soundwalks around the lake as an exercise in active listening and co-creation of place.

Recent years have seen a growing interest in walking—and "walking with"—as a multisensory methodology to perceive and record the hard-surfaced world of our cities through grounded attention to the intricacies of everyday "footwork" (Ingold 2004). As a modal spatial practice, walking enables relational forms of "ambulatory knowing" through movement rather than stationary observation (Ingold 2010, S122), while cultivating care through bodily attunement to place (Altés Arlandis and Lieberman 2021).

Scholars have noted how walking-as-method can generate counter-cartographies that unsettle the "normalizing inheritances" of ambulant research, including histories of settler colonialism (Springgay and Truman 2019, 7), geographies of racism (McKittrick 2011), and assumptions of able-bodiedness (Edwards and Maxwell 2023). Walking extends beyond mere rhythmicity, with each footstep striking the pavement in a steady drumbeat, as described by Solnit above (2001, 3). As Cadogan (2016) argues, racialization, particularly anti-Black racism, disrupts the rhythmic cadence of walking, leading to experiences of place defined by exclusion and the threat of violence, contrary to the emancipatory urban belonging depicted in the literature. Walking thus embodies a privilege not equally afforded to all. Who walks where, how one strolls, when and with whom, all raise critical questions about power, inequality, and rights to the city that pandemic policies of spatial distancing brought to the fore.

The concept of walking-with offers a critical intervention into these dynamics. As an embodied practice with "lived qualities" (Wunderlich 2018), it decenters the human-as-individual to underscore alliances with other-than-human beings and non-sentient things with which space is shared and pathways forged. In the urban ecosystem in my lakeside neighborhood, this included such relational beings and vital things as fish and frogs; plastic and other human-made waste; algae, leaves, and dirt; metal, bamboo, and asphalt; plants and insects, to name but a few of the "lively relationalities" that "intra-act" to constitute threshold landscapes where land and water meet (Barad 2007, 393).

Walking-with not only disrupts human exceptionalism but also calls into question the ocularcentrism that underlies accounts of walking as primarily visual activities (Ingold 2004, 327). Scholars have emphasized other sensory dimensions of walking that decenter vision, particularly the *tactile* sensations

FIGURE 24. Foot travelers on Ngũ Xã (Trúc Bạch) Island, founded in the seventeenth century as a bronze casting village, encountered this child-size cutout of a youth pioneer who reminds them to wear a mask and maintain a minimum distance of one meter. The cordoned-off area behind the figure marks a makeshift gym where young men lifted weights early mornings and late afternoons before the pandemic. Photo by the author.

generated from the force of the foot making contact with the pavement (de Certeau 1984, 97). While walking "incites a touching encounter" through the mechanics of movement (Springgay and Truman 2019, 139), so too does it evoke spontaneous *olfactory* sensations that heighten our awareness of smells, a historically underappreciated facet of sensory life (Corbin 1986, 8). Foul odors, such as the morning stench of stagnant sewage and debris wafting from the canal on my morning jog, acted as powerful signs of infrastructural neglect and the accumulation of urban waste channeled into vulnerable watersheds. Rather than parceling the senses, walking immerses us in overlapping sensory experiences that deepen our connection to urban space.[14]

As Hanoi emerged from lockdown, its sonic composition underwent a dramatic transformation. Soundwalks, involving participant listening-in-motion while attuning to the distinctive acoustic ecology of the city, aligned seamlessly with the conditions of spatial distancing. This methodology holds revolutionary potential; for example, as Black feminist practice, soundwalks legitimize and prioritize Black sonic perception over white normative standards of hearing (Martin 2019). More broadly, this practice shifts attention from the visual ordering of space to an understanding of how changes to soundscapes mediate urban life.

Soundwalking also proved to be potentially restorative, offering insight into the relationship between urban sound and well-being. While extensive research has examined the adverse effects of urban "noise" on health, with quieter cities often regarded as a marker of civilized living (Hsieh 2021, 55), the pandemic revealed the unsettling consequences of noiselessness. Many found the unprecedented stillness disturbing—hardly surprising, given that "a city with sound is a city with life" (Rebuscini and Frassi 2022). The absence of everyday sounds, such as the iconic cries of street vendors, hinted at broader cultural and economic losses for communities. The gradual return of ambient noise, then, signaled the revival of the city—a testament to the state's declared victory over the virus—though this recovery was unevenly felt and heard across the city.

In the following sections, I compare and contrast two soundwalks: a contour walk around the lake and a ghost walk through the Old Quarter (Phố cổ). These walks illustrate disparities in urban revitalization, reflecting different acoustic dynamics of care both within and for the city. My autoethnographic practice of soundwalking moves beyond the conventional view of an individual engaged in deep, passive listening, revealing how bodies that move and perceive collectively create the sensory environments they inhabit. In this way, soundwalking not only allows participation in urban life and its governance but also generates urban liveliness itself through movements that navigate and disrupt its spatial ordering (de Certeau 1984, 98). Walkers, human and nonhuman, are city shapers and place makers of sonic tapestries. They do so tactically and improvisationally, while ascribing diverse meanings to the sounds they encounter and produce within shared spaces.

SOUNDWALK 1: LAKESHORE CONTOURS

In an emergency, the habits of ordinary life may fall away, but other habits come into play, and determine whether the action performed is fatal or benign.
—ELAINE SCARRY, *THINKING IN AN EMERGENCY*

Everyday/Emergency

Early one morning during spatial distancing, before the usual 6 a.m. public announcements, the sound of amplified Latin music woke me with a start. Glancing at my clock, I saw that it was 5:45 a.m. and made my way to the balcony, where the dim morning light revealed a group of dancers gathering on the other side of the lake, conjuring pre-pandemic life. Across the calm body of water, the resplendent bell tower of Cửa Bắc cathedral, exemplary of the hybrid Indochinese architectural style from 1931, stood directly in my line of sight and had roused me even earlier at 5:30. The resonant tolls of the bells had pierced the quiet of early dawn, possibly serving as a call for the dancers to rise and converge at the park. In the distance, I heard a dog barking, followed by roosters crowing in turn. The city stirred, its morning rhythms in sync with the routine soundtrack of the acoustic environment, enveloping me in a deceptive wave of sonic normalcy.

When I arrived at the dance site thirty minutes later—the time it took me to traverse the lake's perimeter—the twirling seniors in their masks and sneakers had become part of a broader sonic ecology, where emergency habits unfolded through sound and movement (Scarry 2011, 14). To the left, serene melodies of bamboo flute and zither accompanied a group seated cross-legged on mats strewn along the paths between the trees, their heads bowed in collective meditation. To the right a lively massage chain of elderly women with outstretched arms—spaced a bit farther apart than usual—followed a recorded voice calling "một, hai, ba . . . [one two three . . .]," gently patting the backs of the people before them in a gesture of mutual, touch-based care (video 3). The fluffy toy poodles that some of the dancers had brought with them ran around playfully, without straying too far. Nearby, the creaking joints of exercise machines revving up again suggested their disuse for a period of time. On the sidewalk, excited shouts resounded from a game of badminton as the birdie flew into the street, just beyond a cage of chattering hens left on the sidewalk under a tree. These seemingly ordinary routines embodied radically transformed ways of being-with and caring-for others in an altered sensory world.

This sonic ecology of interactivity signaled what appeared to be an end to the public health emergency—or perhaps the integration of the exceptional into everyday life as people cautiously resumed relationships recalibrated to risk. Spatial distancing, when observed, marked a distinct moment of pandemic temporality; what was once event-full (outbreaks) now seemed uneasily suspended in a state of eventlessness—not fully in the throes of crisis but not yet entirely

VIDEO 3. Reciprocal care relations through touch, guided by audio instructions: massage chain at the south end of Trúc Bạch Lake, 2020. Video by the author.

To watch this video, scan the QR code with your mobile device or visit
DOI: https://doi.org/10.1525/luminos.249.video3

post-epidemic. A lingering sense of insecurity hung in the air despite the appearance of a disease-free city, reaffirmed by the state's daily virus tally. As geographer Ben Anderson notes, the distinction between the "time of the everyday" and the "time of emergency" flattens in the face of ongoing crisis, leading to a condition where "human life is life lived in uncertainty," a state defined by a temporally ambiguous exceptionality (2016, 177).

In the aftermath of the lockdown, familiar morning sounds—echoes of reconstituted habits—gradually reemerged alongside the routine *loa phường* broadcasts, suggesting an interval in the everyday/emergency dynamic, a lull between a ruptured present and the promise of a restored future (Anderson 2016, 180). Bamboo brooms sweeping away fallen leaves from dusty streets, clanking metal doors rolling open to reveal the interiors of lower floors, and the cacophony of caged birdsong under the canopy of trees all signified the suturing of private life back into the public sphere. This crossing of thresholds formed the basis of my daily practice of walking-with, informed by the notion of contouring. Contours, Springgay and Truman (2019, 141) observe, "edge things." Through this edging process, contours mark thresholds, those liminal spaces or passageways of in-betweenness, which conjoin humans and nonhuman beings and things in novel ecological relationships (2019, 141). Anthropologists have long argued that thresholds embody vital potentiality, or what Victor Turner (1967, 97) famously coined the "realm of pure possibility." Rather than simply establishing boundaries, thresholds blur, transgress, and transform them into fluid zones or frontiers of lively activity, power, and interconnectivity.

In my case of lakeside edges, contour walks marked the transition between land and water, interiors and exteriors, solitude and co-presence, silence and sounding, security and vulnerability. As Springgay and Truman remind us, "*To edge means to move gradually, carefully, or cautiously*" (2019, 141, emphasis in the original). In a pandemic, edging demands a delicate balance of care and attunement among those who share the same potentially contaminated air, even as individuals face disproportionate risks (Kenner 2021, 1115). Soundwalks that "edge" cross the threshold from separation to reintegration, obfuscating the line between the ordinary and the exceptional. This sensory practice of attentive listening while moving attunes us to surfaces, sensations, and the quotidian rhythms of labor and life that are intrinsic to "sensual world making" (Stewart 2011, 446). Taking a contour walk around the lake, with an ear to its sonic ecologies of place, allowed me to reconnect with the sensory qualities of urban public space, while remaining mindful that edges can also expose conditions of heightened uncertainty—of lives unsettled, on the precipice of precarity (Saguin 2022, 10).

Sonic Anarchiving: Critical Cartographies of Sound Mapping

As I walked, listened, and sensed that edgy urban world around me, I co-created a sound map along with an auditory *anarchive* of the lakefront's morning sounds to analyze the sonic uniqueness of threshold spaces (figure 25). Yet mapping—even in sonic form—is not a neutral act. Cartographic practices have been deeply entwined with imperial histories of expansion and dispossession through violent or deceptive means. Maps, including sound maps, depict and shape relations of power and claims to territory and its resources (Rose-Redwood et al. 2020), while erasing Black, Indigenous, and other subjugated geographies (McKittrick 2011). My efforts at mapping and recording sound events—or what might be considered a form of *sonic anarchiving*—were not intended for ethnographic preservation or the salvaging of disappearing sounds by a "white female songcatcher" performing the task of savior and custodian of intangible heritage (Kheshti 2015, 19–20).[15] Rather, my motivation drew inspiration from the field of critical cartography, especially its application in Vietnam (Kim 2015). I sought to hear, identify, and amplify edgy soundmarks, those sonic-spatial activities typically excluded from conventional mapping, which hold the potential to challenge the power of the state to determine the use and sounding of public space.

Guided by an "anarchival impulse" (Foster 2004, 5), my shift from landmarking to soundmarking prioritized obscure and ephemeral traces over absolute origins and immutable sources to rethink established hierarchies of sonic dominance. I aimed to chart alternative geographies of power and vulnerability, recognizing

FIGURE 25. Sound mapping: walking path marked with select auditory activities, comprising the lakeside's sonic ecology. (X: group exercise; M: meditation groups; F: fishing; O: outdoor fitness equipment; S: team sports; V: vendors; and B: birding.) Source: Google Earth.

that urban lakes serve not only as resources for recreation but also for livelihood creation. My attention centered on mundane sonic events often overlooked by more spectacular approaches to archiving, which typically feature conspicuous elements like musical practices worthy of designation as "heritage." A sonic anarchive functions as an auditory *"repertoire of traces"*—recurring sounds and acoustic signatures that do more than merely document past encounters but also operate as a *"feed-forward mechanism"* that amplifies evolving creative processes (Massumi 2016, 6; emphasis in the original). In other words, anarchives function as dynamic—potentially anarchic—processual starting points rather than static, taxidermic endings.

Sonic ecologies are ever mutating. The state-dominated soundscape of lockdown described in act 2 gradually yielded to a more dynamic acoustic environment, produced collectively through human and nonhuman interactions, which I capture in the above sound map. While routine sounds may create a sense of comforting familiarity, how individuals perceive, understand, and interpret shared sensations—be it heat, smells, sounds, or a breeze—remains culturally and historically contingent. Perceptions of "noise" are deeply rooted in specific social and spatiotemporal contexts, intertwined with existing relations of power and privilege (Hsieh 2021, 54). Individual sensitivities to sonic events vary widely, and assumptions about sound can be used to mark differences—even presumed levels of civilization—through culturally reductive binaries. For example, generalizations that characterize Westerners as noise-sensitive and Vietnamese as exceptionally noise-tolerant tend to break down under close ethnographic scrutiny, as sonic anarchiving reveals the unique and unpredictable ways people produce, interpret, and negotiate sounds differently across and within their soundscapes.

Consider the varied reactions to public loudspeakers discussed in the previous chapter, which illustrate a generational critique of state power, disguised as a physiological response to intrusive noise. Urban youth, prone to questioning authority, criticized the outdated technology for disrupting their sleep, while early risers, sentimental about these media devices, valued the timely delivery to information as evidence of effective governance. This example highlights the absence of a universal perception or cultural interpretation of sound. The diverse meanings we ascribe to the sonic world around us provide insights into both listeners and sound producers entangled within a complex sensory network of power-laden constructs. Sonic anarchiving thus proves essential to analyzing these multifaceted, sound-mediated relationships.

Acoustic Ecologies: Pandemic Encounters with Nonhumans

The pandemic-induced "anthropause," a term coined to identify the global reduction in human activity, sparked widespread discussion about the resurgence of urban wildlife. This discourse often centered on eco-acoustic transformations to cities, where quieter urban atmospheres amplified the sounds of nature. A multispecies study published by *Science* in September 2020, for instance, found that the reduction of human-generated sounds, or what scholars have identified as "anthrophony," during citywide shutdowns altered the frequency and intensity of sparrow singing in vacated spaces (Derryberry et al. 2020). Such findings resonated with me, given my own attunement practices and understanding of how noise—and its absence—shaped my affective state of being. During the pandemic I found birdsong to be particularly soothing and anxiety reducing. However, unlike the scenario described in the *Science* study, birds did not flock to my neighborhood during Hanoi's "silent spring." In fact, the decrease in avian sounds, real or otherwise (such as the rhythmic splashing of swan pedal boats gliding across the lake on weekends), contributed to the eerie quietude that befell the city. This was largely because the lively avian soundscapes I encountered during my morning soundwalks mostly originated from *caged* birds, with some, like roosters, engaging

AUDIO CLIP 5. Nonhuman sonic ecology: interspecies dialogue during early morning soundwalk around Trúc Bạch Lake. Source: author.

To hear this audio clip, scan the QR code with your mobile device or visit DOI: https://doi.org/10.1525/luminos.249.audio5

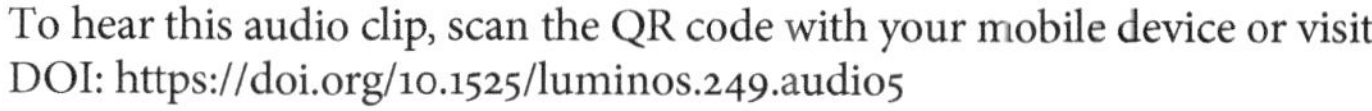

in vocal communication across distances, from one street or house to the next (audio clip 5). During the lockdown, these nonhuman city-dwellers were either brought indoors or relocated to the countryside by families fleeing for what they perceived to be safer, virus-free ground.

In Vietnam, the hobby of keeping and caring for birds is predominantly a male activity. Along my lakeshore contour walks, I regularly observed men who displayed affection toward their feathered pets: some groom their game fowl (in preparation for cockfights, a well-known—though banned—pastime among some men in Trúc Bạch Ward), while others house their songbirds in ornate bamboo cages, carrying them outside to be displayed in the early morning hours. The bird trade is a highly lucrative industry, with native species dominating the market, according to wildlife monitoring networks.[16] The value of a bird is pegged to its vocal abilities and singing repertoire, one of my interlocutors explained while mimicking various bird vocalizations. Birdsong is not only enjoyable to listen to, he told me, but it also serves a deeper cultural purpose. In the home, it contributes to positive *phong thủy*, or feng shui, creating *đất lành*, a harmonious living environment conducive to a peaceful and prosperous life. The cultural significance of birds is exemplified by the proverb "Đất lành chim đậu," or "On auspicious land, birds will perch," which implies that the presence of melodic birdsong signals good fortune to come. Conversely, it also suggests that one should refrain from building a house or establishing a grave (a "home" for the dead) on land devoid of an active, chirping avian community.

The melodies of canaries and finches welcomed me as I stepped out of the house for my "reintegration" walk that April morning after the shelter-in-place order was lifted (video 4). I came upon my neighbor, a retiree in his seventies, who had come out of his home and was crossing the street with his birds. I watched as he used a long bamboo pole to suspend their cages from the electric line that ran alongside the lake, sheltered by the trees. The birds seemed excited to be outdoors again; they chirped among themselves and jumped about their cages. This resurgence of life and sound unexpectedly filled me with a sense of hope, adding a spring to my step as I picked up my pace. I rounded the bend, past the thwacking punts of *đá cầu* or foot badminton, Vietnam's "street sport," and crossed the canal to the eastern side of the lake, holding my breath to avoid the stench. I made a sharp left and, glancing to the right, noticed a vacant plot, where construction had halted.[17] From there, the synchronized crowing of a pair of caged roosters,

VIDEO 4. Return of the nonhumans: birdsong around Trúc Bạch Lake became a vibrant, almost therapeutic, feature of the soundscape during the period of spatial distancing, 2020. Video by the author.

To watch this video, scan the QR code with your mobile device or visit
DOI: https://doi.org/10.1525/luminos.249.video04

under the care of the *người bảo vệ*, caught my ear. Down the way, I encountered a group of chattering, free-roaming hens and their chicks in the road. They strolled alongside me briefly—as sidewalks in Hanoi typically serve purposes other than walking (Nguyễn 2022)—before pivoting after a few meters to head back to their coop, nestled in a lush, lakeside garden.

While contour walking facilitated interactions with sentient, nonhuman beings and their unique vocalizations, not all eco-acoustic encounters were positive or pleasurable. As sound ecologist and pioneering soundwalker Hildegard Westerkamp (2017) reflected on the radical possibilities of listening in unsettling times: "The seemingly simple act of listening to the environment often leads to unexpected complexities of thoughts, sensations and emotions that are not always comfortable." Farther along my walk, past the outdoor fitness equipment and over the perimeter fence, I came upon a scene largely hidden from the eyes and ears of the state. On the stone embankment, fisherfolk deftly maneuvered their bamboo poles, creating ripples in the water as they hauled in their catch and dropped them in a growing pile. Like birding, fishing is considered a masculine pursuit, serving as a leisurely pastime for some and a source of livelihood for others, operating at times on the margins of legality.[18] The subtle sounds of their activity—the soft splash of lines cast and fish pulled from their habitat—added another layer to the area's rebounding acoustic landscape.

The origins of Trúc Bạch Lake, separated from the much larger West Lake by a dike constructed in the seventeenth century, can be traced to aquaculture. For generations, these lakes have been vital sources of sustenance and income, despite unauthorized fishing in West Lake being prohibited and regulated since 1958.[19] At Trúc Bạch, individual operators often escape the notice of authorities, who direct their efforts toward monitoring commercial poaching in the more fish-rich waters of West Lake, Hanoi's largest freshwater body.[20] The carp in Trúc Bạch Lake are comparatively smaller, and recent fish kills from oxygen depletion have prompted ward officials to install fountains around its perimeter. These aeration devices not only improve the ecosystem but also add the soothing sounds of cascading water to the shoreline. I approached one fisherman as he reeled in his modest carp and asked him, "How much have you caught today?" "Three kilograms!" he responded proudly. He explained that he planned to take the fish home to eat rather than sell them to vendors in the nearby market or adjacent hot pot shops—places I often observed people peddling their catches. "Do you want to buy some?" he asked, with anticipation in his voice.

Nearby, a woman squatted beside a winnowing basket, methodically snipping off the heads of sardines with her rusty scissors—snip, snip, snip. This was just one of the many (for me, less comforting) sounds associated with food preparation, which, in this case, signaled the hopeful reopening of her small food operation. On the main strip (the former dike), fisherfolk tossed their morning catch onto the cracked pavement. The sight and sound of fish flopping—really, asphyxiating—

with some reaching great heights while others lay inert, gazing skyward, served as a poignant reminder of our status as sentient, relational beings. I recognized, however, that my sensitivity to this scene likely diverged from those of other passersby. Melodic birdsong—and avian soundscapes more broadly—had instilled in me a momentary surge of optimism and tranquility. In stark contrast, the sounds of fins flapping and gilled bodies smacking against the concrete stirred feelings of distress at what I perceived to be the violence of their slow death. For me, this became an allegory for worldwide suffering, evoking the desperate struggle for oxygen among those afflicted with COVID-19. Yet I understood that these same human-animal interactions might signify something entirely different to others, something ordinary or perhaps even reassuring, like food security in precarious times. As Rebecca Solnit notes, walking "can be invested with wildly different cultural meanings" (2001, 3); the same holds true for the sounds we encounter along the way.

Anthrophonics of Lakeshore Contours

Soundscape ecologists have proposed a tripartite framework for understanding the spectrum and spatiotemporal patterns of sounds that emanate from the environment. This framework consists of *biophony*—sounds produced by nonhuman beings, such as the birds and fish I described above; *geophony*—ambient sounds from non-living nature like wind and rain; and *anthrophony*—sounds generated by humans and their technologies (Pijanowski et al. 2011). While these categories are useful for analyzing soundscapes, it is important to recognize their relational nature as overlapping, evolving, and interdependent.

During the pandemic in Hanoi, the distribution of sound among these "phonies" shifted significantly. The sharp reduction in human-generated noise (anthrophony) allowed other sonic phenomena to become more audibly prominent. These modifications to the acoustic environment highlighted the sonic synergies among these categories and how a change in one could profoundly alter perception of the others. The following three examples of what I call "lakeshore anthrophonies" demonstrate how this rebalancing of sounds manifested in the everyday actions of state and non-state actors following the nationwide lockdown.

Lakeshore Anthrophony 1: Sidewalk Rhythms of Regulation   As spatial distancing measures took effect, the city's sonic ecology once again transformed, showing how pandemic governance reverberated through urban sound. Although biophonic activity, such as birdsong, increased noticeably, the resurgence of anthropogenic sound occurred more incrementally. Traffic noise remained subdued, while restrictions on street vending, which manifested in a variety of trading practices around the lake, left certain shoreline spaces where people once gathered quiet and vacant. In non-crisis moments, the anthrophony of lakeside contours was dominated by sounds from both leisure and livelihood activities. A prime example was the spontaneous "frog market" (*chợ cóc*) on the eastern edge of the lake, situated

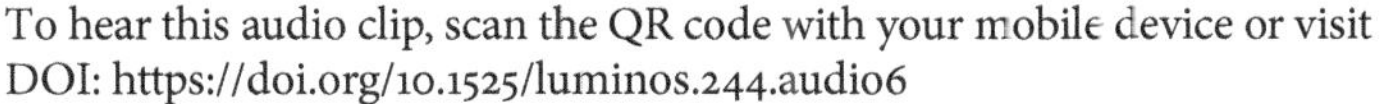

AUDIO CLIP 6. Mobile police loudspeaker unit: patrolling sidewalks along Trúc Bạch Lake between health broadcasts during the pandemic. Source: author.

To hear this audio clip, scan the QR code with your mobile device or visit DOI: https://doi.org/10.1525/luminos.244.audio6

just past the fish and hens, before one turned onto the bustling strip of Thanh Niên Road. This early morning trading spot evolved from humble beginnings with only a few female migrants from the countryside selling vegetables from the backs of their bicycles at Trúc Lạc Alley. Over time, it mushroomed into a boisterous flea market of sorts, with used clothing and household items spread across plastic tarps that appropriated the entirety of the pavement.

There was a distinct temporality to this lively sunrise soundscape. The market grew organically, attracting more traders as it expanded around the corner, eventually stretching down the walkway to Thủy Trung Tiên Temple.[21] By 7 a.m. however, this busy hub of commercial activity would vanish abruptly, just before the traffic police started their first round of sidewalk checks. The ebb and flow of daybreak sounds reflected a delicate balance between mutual care and livelihood risk, as vendors and business owners established an informal warning system to alert one another to approaching patrols that might confiscate their goods.[22] At times, police showed sympathy through their own sonic ritual, a series of short loud honks signaling their imminent arrival. This auditory cue would spur people into frenzied action, followed by a bullhorn announcement instructing individuals to clear any items encroaching on public space (audio clip 6). Once the patrol had passed, commercial and social activities would resume until the next round began.[23]

The corner frog market would not continue operations after the lockdown; authorities were able to use the pretense of the pandemic to enforce "sidewalk cleanup" campaigns through increased policing for "public space violations." Yet the female vendors on their bicycles, who could pedal away quickly, did eventually return to the city. Their reappearance near Trúc Lạc Alley hinted at the gradual resurgence of the informal economy in my neighborhood marked by an increase in human acoustic activity. Green tea vendors, for example, were back in their usual locations by May 1, just a week after the stay-at-home order ended. Along the main stretch of Thanh Niên Road near the John McCain monument, makeshift tea stalls run by women were popular gathering spots for older men. With their well-groomed—and, in winter, sweater-clad—dogs in tow, these men would settle onto low plastic stools along the walkway, socializing and sipping tea before the morning sun climbed too high. This lively scene illustrated how soundscapes could be gendered according to spatiotemporal variability. In this particular space along the contours of the lake, during my sunrise soundwalks, male voices dominated the anthrophonic sounds, punctuated by the biophonic barking of their dogs.

A short distance away, at the northwest corner of the lake, just beyond a group of salsa dancers and occasional yoga practitioners marking the halfway point of my route, I sometimes paused for a break. There, a kind-faced woman in her seventies would serve me a glass of iced tea under the canopy of trees she had planted thirty years ago when she first set up her stall.[24] During these moments, she often shared her stories with me, including how authorities had prohibited her from vending during April 2020.[25] Slowly, the adjacent soup stands and cafés began to reopen, places where she had heated water when her makeshift setup, tapped into power lines, failed. Their seating—occupied but widely spaced—reflected the cautious resumption of urban acoustic life, as people remained alert both to visual and audible signs of ill health, and to impending sidewalk inspections.

Despite repeated warnings from public loudspeakers and roving sound trucks to "limit going to public places and gathering in crowded areas" (Hạn chế đến nơi công cộng, tập trung đông người), the vitality of urban life proved resilient. Lakeside cafés, tea stalls, and public parks continued to attract people seeking social interaction and outdoor recreation. The anticipation of a return to urban normalcy was further reinforced at the start of May by a mini marathon around the lake. This event appeared to challenge current restrictions: unmasked runners in close proximity, perspiring and breathing heavily, releasing droplets of respiratory fluids into the atmosphere as their feet thudded rhythmically against the pavement while observers cheered. This festive scene vividly demonstrated how state efforts to control urban spaces through sound and surveillance ultimately gave way to people's determination to reclaim their pre-pandemic ways of life.

Lakeshore Anthrophony 2: Sonic Trade Perhaps no anthrophonic sound was more suggestive of urban revitalization than the melodies of itinerant street vendors, whose calls were brought to life and amplified by technology. Street vendors and their distinctive cries are a pervasive feature of Hanoi's urban soundscape. In my lakeshore neighborhood, a hawker would pass by my house every twenty to thirty minutes. From my seventh story flat, I could sometimes hear them calling out simultaneously, their voices carrying from both the street in front and the alley in back. It comes as no surprise, then, that migrant women traders have become one of the most studied subjects in urban studies of Vietnam. And yet remarkably little consideration has been given to the sonic dimensions of their labor. This is a curious oversight considering that for many, the volume of their sales hinges on the pitch of their voice and the range it can carry, with the goal to entice people out of their homes or cafés to purchase the goods they desire.

When the pandemic hit, the government's ban on street commerce forced vendors to return to their home provinces, silencing their familiar calls that once filled the narrow streets. After travel restrictions eased, migrant traders gradually

AUDIO CLIP 7. Shifting vocalizations: an itinerant *tào phớ* (tofu pudding) vendor alternates between live and prerecorded calls while navigating street acoustics. Source: author.

To hear this audio clip, scan the QR code with your mobile device or visit DOI: https://doi.org/10.1525/luminos.249.audio7

00:00:16 ——————————— • ——————————————————————————— 00:00:46

Audio clip 7

returned to the city, their voices once again animating urban spaces, but now accompanied by sound reproduction technologies that intensified during the period of social distancing.

Street vendors' distinctive cries play a vital yet often overlooked role in the formation of urban media ecologies. Their methods of communication, increasingly reliant on do-it-yourself (DIY) audio technologies rather than traditional vocalizations, illustrate the constant evolution of media infrastructures. As Mattern (2015, 104) notes, these infrastructural constellations are produced not only by institutional forces but also by the everyday practices of ordinary people. This insight proved particularly relevant during social distancing, as renewed sonic connections diversified the soundscape beyond the state-dominated acoustic environment of lockdown.

The shift toward audio technologies among street vendors has sparked debates about modernity's impact on urban soundscapes, particularly regarding the separation of sound from its human source. Critics often romanticize what they see as the "loss" of authentic singing to prerecorded audio, viewing traditional calls as endangered sonic heritage. These nostalgic perspectives risk casting migrant traders as timeless figures unaffected by modernization, preserving what is imagined as a more authentic—and less mediated—sonic ecology of the city. This tendency aligns with Nina Sun Eidsheim's observation that the true source of a voice lies not in the singer but in the *listener* (2019, 9), whose cultural assumptions about the vocalist—in this case, someone untouched by technology—shape their perception of the sound.

In my ward, however, street vendor technology—battery- or bicycle-powered speakers—exemplified a pattern of intertwining systems, or sonic convergences, rather than a simple replacement of human voice. This mirrors my argument about public loudspeakers in act 2. A prime example is the tofu pudding (*tào phớ*) vendor in Trúc Bạch Ward, who returned to the streets in late May 2020. This vendor stood out as one of the few male hawkers in my neighborhood, and the only one to utilize both live singing and taped recordings.[26] His audio choices responded to the urban acoustics of the material environment. In the narrow alleys, where his voice echoed between buildings, he called with a distinctive musicality, "Here is tofu pudding! Who wants tofu pudding?"[27] However, when he merged with traffic on busier and louder streets, he adapted by switching to a speaker system powered by a battery attached to his weathered bicycle, allowing him to compete with the clamor of the city (audio clip 7).

FIGURE 26. The return of itinerant vendors to Trúc Bạch Lake. This *bánh đa kê* (rice millet cracker) seller ingeniously rigged a portable speaker system, fueled by a pedal-operated battery concealed in the plastic bottle, to broadcast a prerecorded call for her popular snack through an amplifier mounted on the handlebars, May 2020. Photo by the author.

AUDIO CLIP 8. Kinetic-powered amplification: the *bánh đa kê* vendor's mobile speaker system. Source: author.

To hear this audio clip, scan the QR code with your mobile device or visit DOI: https://doi.org/10.1525/luminos.244.audio8

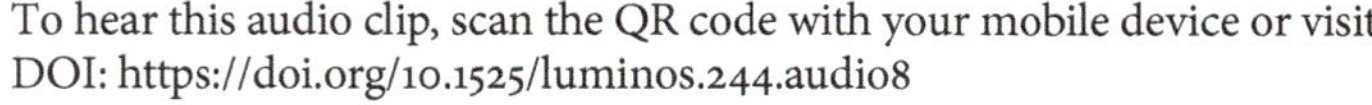

00:00:02   00:00:18

Audio clip 8

In other instances, devices connected to cart or bicycle wheels converted rotational force into energy to charge batteries. This kinetic energy harvesting system created a dynamic synergy between speaker and vendor, with the speaker's power entirely dependent on the vendor's pedaling movement. When the vehicle came to a halt, such as when making a sale, so too did the speaker stop its sonorous call. During the pandemic, the use of such improvised technology with looped recordings ("Who needs a broom?"; "Who wants sandals?") helped reduce the risk of airborne transmission by minimizing the release of respiratory droplets during calls and transactions (figure 26; audio clip 8).

FIGURE 27. Gendered soundscapes: on the left, a popular fitness instructor combines leisure and livelihood with a motorbike-mounted speaker; on the right, the Ward People's Committee sets up a PA system for morning exercises, extending the state's sonic presence through diverse audio media. Photos by the author.

As these technological adaptations became more prevalent, the once familiar echo of human voices grew rarer. Although practical and safer, these automated calls transformed the acoustic texture of urban life into something that felt more mechanized and sounded less sonically intimate.

Lakeshore Anthrophony 3: Cardio Beats   The assemblage of street vendor and state authority voices, both live and recorded, underscored the critical role of audio technologies in co-creating the lakeside's human-generated soundscape during spatial distancing, alongside interactions with nonhuman sounds. Beyond vending, leisure activities also revitalized the acoustic environment post-lockdown, often with gender-specific patterns. While tea stalls served as hubs for male sociality, outdoor fitness classes were popular gathering spaces for women, their sound systems often drowning out public health service announcements. During morning soundwalks, I encountered no fewer than six exercise clubs around the lake, each with distinct demographics, styles of dress, and workout routines, showcasing a range of urban subjectivities. Each group had a unique sonic profile, amplified by portable speakers brought by organizers: either transported by motorbikes or rolled out from ward cultural centers, showing a mix of independent initiatives and state-sponsored programming (figure 27). Their eclectic, high-volume playlists varied daily, featuring pandemic hits like "Hey Vietnam!" as well as salsa

VIDEO 5. Early morning exercise groups promote collective health while strengthening social bonds along lakeshore contours, adapting fitness routines to social distancing guidelines, 2021. Video by the author.

To watch this video, scan the QR code with your mobile device or visit
DOI: https://doi.org/10.1525/luminos.249.video5

mixes, throwbacks from ABBA or Modern Talking, folk tunes with instructional voiceovers, wartime ballads, and, most frequently, Vietnamese pop music, or V-Pop. In this way, women played an important role in transforming lakeshore contours into fitness zones through sound and a shared sense of sonic belonging. Outdoor exercise emerged not only as a form of healthful physical activity but also as a response to the isolation of lockdown, fostering collective care and well-being in public spaces where social connection was otherwise restricted.

Fitness groups formed synchronized corporeal communities around sonic events that presented a "dangerous intimacy" amid an ongoing epidemiological threat (Jones 2020). Participants navigated these risks while adhering to public health guidelines that framed exercise (*tập thể dục*) as essential for healthful and hygienic living—collaborating with state sonic authority rather than opposing it (video 5). This dual nature of fitness aligned with broader COVID-19 messaging that reminded the public of their duty to safeguard society, making personal well-being inseparable from national health security. Far from embodying neoliberal notions of the self as an autonomous manager of individual responsibility and self-care, fitness practices reflected an understanding of subjectivity as relational and rooted in shared vulnerabilities.

Vietnam's relational approach to health has its roots in the historical development of sports to advance nation building, dating back to its anticolonial resistance. This history of exercise—including walking—reveals how fitness and sports have been used as biopolitical tools to improve health, cultivate solidarity, and regulate bodies to realize state objectives, like revolution and the building of socialism. Exercise, in other words, was embraced as a path to emancipation. Hồ Chí Minh's 1946 movement, "Khỏe vì nước," or "Health for the Country," embodied the moral and spiritual values of "revolutionary physical education and sports" (*thể dục, thể thao cách mạng*) that called on people nationwide to engage in exercise for purposes of national defense and development of a shared cultural identity.[28] Under the slogan, "Dân cường thì quốc thịnh," or "A strong people make for a prosperous country," the welfare of every citizen was considered vital to the progress of the nation, rather than to individual flourishing alone.[29]

The Ministry of Health's emphasis on exercise to maintain health and foster social cohesion for the purpose of biosecurity exemplified the principles of sonic socialism during the period of spatial distancing. Around Trúc Bạch Lake, morning sport and exercise enthusiasts aspired not only to build strong, fit bodies, but also to nurture social relationships, while reinforcing national identity through sonic alignment with the state's pandemic response. This included, for example, incorporating "Việt Nam Ơi!" into cardio routines. Fitness groups also publicly expressed their allegiances on national holidays by coordinating their clothing and music. For instance, they wore red shirts on Independence Day (September 2) and green shirts on Vietnamese Women's Day (October 20), while syncing their soundtrack to the occasion, such as playing patriotic songs during national celebrations. After these special workouts, women lingered at the lake, sitting together on bamboo mats to share food and converse, at times inviting me to join the festivities. This exemplified the interconnectedness between autonomous bodies and the broader body politic. In public spaces, practices of self-cultivation (exercise) and self-regulation (spatial distancing) mirrored the anxieties of a society navigating a public health crisis. These practices drew on socialist genealogies of disaster management that mobilized multisensory tools, especially sonic ones, to frame care as a shared obligation, a collective act of mutual responsibility for the common good.

SOUNDWALK 2: GHOST TOWN

If an [abandoned] building could speak, what would it say? What would it sound like?

—DAVID LITTLEFIELD AND SASKIA LEWIS, *ARCHITECTURAL VOICES: LISTENING TO OLD BUILDINGS*

The First Death: From Boom to Doom

In early May, two weeks after the last reported case of community transmission, the government declared that the coronavirus had been successfully contained and brought under control. A week later on May 8, the Ministry of Culture, Sports, and Tourism announced the launch of a new initiative: Plan 1749/KH-BVHTTDL entitled "Người Việt Nam đi du lịch Việt Nam," or

"Vietnamese people travel Vietnam." This stimulus program sought to promote mass domestic tourism in the context of a "new post-pandemic normal" (*bối cảnh bình thường mới sau đại dịch*). Plan 1749 focused on reopening hotels, airports, and attractions across the country with the goal to stimulate the economy and offset the loss of international tourism by reducing travel costs and fees while maintaining service quality and public health safety.

This initiative proved surprisingly successful. As international borders remained closed, tourism within Vietnam flourished.[30] Emerging forms of middle-class mobility beyond lakeside activities like walking, fitness, or biking showed that people were cautiously venturing outside their comfort zones to navigate new horizons of risk. Steep discounts—up to 60 percent for luxury establishments in some cases—led to a surge in bookings, as the hospitality industry redirected its efforts toward attracting a domestic market. With schools shuttered, family promotion packages enabled some children to spend their days not in classrooms but in pools. The cheerful sounds of their lively splashing and laughing were yet another indication of the atmosphere of normalcy that seemed to envelop Vietnam. Within Hanoi, the concept of "staycations" (*du lịch tại chỗ*) gained traction among Vietnamese eager for adventure but cautious of traveling too far from home due to the risk of *in situ* quarantine. Hotels amplified this blend of ordinary life in extraordinary times by hosting evening musical performances that featured Filipino ensembles on the hotel circuit, making visible the presence of stranded migrant workers unable to return to their home countries (see pandemic figure 5).

This caution was not unfounded. On July 25, as the Ministry of Health marked the ninety-ninth day without community transmission, an outbreak in Đà Nẵng and the discovery of a mutated viral strain shattered the country's record, leading to its first COVID-19-related death one week later, a moment that would soon be eventalized.[31] At the time, my spouse and I were celebrating my birthday in Sa Pa District, Lào Cai Province, a place where "people did not pay much attention to the [April] lockdown," a Hmong farmer tending her hillside garden told me, because the virus was too far away. This woman perceived risk as originating from outside the region, consistent with the construct of the "stranger" as microbial threat. In this case, however, it was not foreigners but dominant Kinh lowlanders who introduced a state of "bioinsecurity," acting as conduits of a disease that "threaten[ed] to disrupt the fabric of life" (Ahuja 2016, 197).

News about Đà Nẵng reached me through a series of rapid pings to my phone, prompting us and the Vietnamese families staying at the highland eco-lodge to hurry back to Hanoi. As panic spread, the central government, in coordination with local authorities, orchestrated an immediate, mass evacuation of tourists from the coastal city, activating the mechanisms of disaster socialism. An estimated eighty thousand visitors enjoying their summer holidays on crowded sandy beaches, including thousands from Hanoi who had driven to the central region to avoid mass transportation, fled before the city locked down three days later on July

28. This abruptly halted domestic tourism, dashing hopes of a prolonged return of normalcy.[32]

As the perception of Vietnam as a virus-free haven came to an end, so too did the practice of using *"hậu"* (post) to refer to COVID-19. The resurgence of the coronavirus, marked by the first fatalities, shattered the optimistic outlooks for economic recovery. As infections spread from beachgoers to factory workers, various industries suspended operations. Businesses shut their doors once more, leading to higher unemployment and greater precarity. The Asian Development Bank (ADB) revised its 2020 growth forecast for Vietnam from 4.1 percent to 1.8 percent. This downgrade occurred even as Vietnam gained global recognition as one of the best performers in containing COVID-19 and one of the first countries to fully reopen its domestic economy.[33]

This mantra of "best performance"—whether pertaining to the state of the economy or government containment practices—offered only a partial view of a more complex pandemic landscape.[34] The economic impact of COVID-19 was unevenly distributed across the city. While certain sectors thrived, particularly family-run, open-air establishments in my neighborhood that offered outdoor spaces to congregate, others languished. This was especially the case for less adaptable businesses that relied on international tourism as their primary source of revenue. This patchy geography of revitalization was reflected in Hanoi's shifting soundscapes and their variable acoustic intensities. Previously, I described how threshold spaces experienced diversified anthropogenic and biophonic sounds, revealing social and sonic investments in care-full relations between humans, nonhumans, and the state. However, not all urban areas flourished or received care equally. Theorists of care in the city have argued that the pandemic exposed an urban paradox: while some spaces benefitted from attentive caregiving, others suffered from neglect or *uncare*, marked by emptiness, abandonment, and, I would add, stillness (Gabauer et al. 2021, xviii). In those areas, the "new normal" manifested not as revival but as rupture, evoking an uncanny sense of urban spectrality. This sensation of care as estrangement, which drew attention to the scale of the pandemic's impact, bore an eerie resemblance to that of a ghost city.[35]

Urban Specters

The notion of a "ghost city"—and urban ghosts—has been widely discussed in scholarship and mass media as a metaphor for the urban decline, neglect, absence, and ruin intrinsic to racial capitalism.[36] Here, I employ this concept to allude to the pandemic's disruptive potential in urban neighborhoods where their intended purpose—in this case, socioeconomic activity—ceased to exist, creating precarity attributable to global capitalism. Like ecological disasters (e.g., Chernobyl), deindustrialization (former mining towns), dispossession (abandoned settler colonies), and over-development (Chinese "ghost cities"), pandemics generate downturns and disparities. These human-made disasters interrupt the urban

VIDEO 6. Once a vibrant hub, Tạ Hiện Beer Street echoes with the solitary clanging of a trash collection bell, a stark reminder of the disproportionate risks faced by essential workers who maintain urban infrastructure, 2020. Video by the author.

To watch this video, scan the QR code with your mobile device or visit
DOI: https://doi.org/10.1525/luminos.249.video6

rhythms of everyday life, leaving hauntings in their wake. This was especially true in sections of the Old Quarter that catered to international travelers, such as the Tạ Hiện "Beer Street" neighborhood, once popular with younger, budget-conscious backpackers (video 6).

Ghost towns signify temporal disruptions—ruptures to collectivities, ecologies, lives, and livelihoods left to linger in an atmosphere of emptiness, silence, and suspension. The material remains of deserted localities evoke a haunting sense of place, shaped not by specters of the colonial past but by unrealized futures. Postcolonial ghosts and ghost towns, in particular, signify threshold spaces on the precipice between presence and absence, material and immaterial, liveliness and death.[37] As such, they help us make sense of the pandemic's "exceptional eventfulness" (Roth 2020), experienced by many as inequality and injustice. Ghosts and ghostliness, in other words, tell us something about the Anthropocene and contemporary conditions of humanity through the looming specter of a disproportionately impactful disease.

Listening across urban sonic ecologies during the pandemic not only generated distinct senses of time and place but also exposed deep disparities throughout the city. Attuned listening revealed dramatic disjunctures between adjacent neighborhoods, yielding vastly different sensory experiences of urban space. The flourishing city coming back to life uneasily coexisted with the desolate city of the departed. While the ghost town atmosphere of Tạ Hiện Beer Street persisted, just a few blocks away, the lively Hàng Mã (Votive Paper Street) buzzed with customers preparing for the annual hungry ghost and mid-autumn festivals, only weeks after the consequential Đà Nẵng outbreak. These stark acoustic contrasts were palpable during my soundwalks. Around Trúc Bạch Lake, the bustle of noise and social activity evoked a strong sense of recuperation and reconnection, while the ghostly silence of the Old Quarter revealed an unsettling sonic void, suggestive of disconnection and decline.

Ghost towns, like cemeteries, attract walkers—myself included—with their evocative surfaces, acoustic resonances, and charged atmospheres. As places of transition suspended between what was and what may (again) become, they are infused with an air of desolation, while hinting at latent possibilities (Springgay and Truman 2019, 138). On my soundwalks, Tạ Hiện Beer Street and its surrounding vicinity, including the nearby historic (and then-closed) Đồng Xuân marketplace, embodied these affectively rich, threshold qualities—haunting reminders of pre-pandemic life, imbued with potential for renewal.

My sensory walkabout began at the intersection of Lương Ngọc Quyến, in front of a colonial-era corner building that had previously housed the street's most popular gathering spot. I proceeded down Tạ Hiện toward the barricaded market, a short stretch ending at Hàng Buồm. Shuttered storefronts flanked me on both sides. The once familiar scene of plastic stools and tables spilling into the street, laden with glasses of cheap beer and plates of snack food, had vanished. The boisterous foreign tourists who had previously packed the narrow alley, reveling into the night, were now conspicuously absent. This newfound serenity allowed me to stroll at a leisurely pace, appreciating the colonial facades and historical "tube house" design in a way that had not been possible before. I turned right on Hàng Buồm and merged onto the now quiet Mã Mây, which curved back around to run parallel with Tạ Hiện. Heading toward Hoàn Kiếm Lake in the heart of the city, I ended my walk at Hàng Bạc, or Silver Street, passing rows of boarded-up shops, restaurants, hotels, and tour agencies, and encountering but a handful of people along the way.[38]

This section of the Old Quarter, normally one of the most densely packed areas in the city if not the entire country, now exuded an uncanny feeling of emptiness and solitude. Even motorbikes avoided shortcuts through these ghostly side streets, opting instead for the main thoroughfares. Vacant roads and buildings in Vietnam are known to serve as magnets for ghosts—lonely, wandering souls—whose presence could bring misfortune unless properly appeased. And yet in this deserted neighborhood, there was no indication of any care given to these urban spirits.

As I honed my senses to detect subtler sonic frequencies, the stillness of the space "spoke" to me, conveying a powerful message about disruption. Just as we can "listen" to images (Campt 2017), we can listen attentively to these other-than-human entities—ghosts, abandoned buildings, or other figures of absence—to discern what they convey about the "quiet soundings" of threshold places (2017, 9). In this particular sensory atmosphere of crisis, these soundings amplified conditions of precarity and neglect as "spectral uncertainty" (Rachwał 2017), diverging sharply from state narratives of pandemic care and success. Quietude then "registers sonically, as a level of intensity" (Campt 2017, 6), drawing attention to the urgency of careful inquiry—an inquiry full of care. This approach allows access to the affective registers through which nonhuman actors "enunciate alternative accounts" (2017, 5), opening up possibilities for speculative, anarchival practices.[39]

If we accept that buildings have voices worthy of our regard, as Littlefield and Lewis (2007) have argued, then listening to shuttered buildings and their ghostly resonances ascribed a certain agency to architecture in a moment of sonic rupture, when the pandemic had rendered their voices silent.[40] This muted and forsaken city of specters bumped up against a more dynamic city of recuperation. Hàng Đường (Sugar Street), the main north-south thoroughfare, roughly divides the Old Quarter into two sections, a western area supported primarily by domestic trade and an eastern part sustained by international tourism. Expanding the range of my soundwalk to gauge the unevenness of the city's recovery, I crossed Hàng Đường from the ghost town of Tạ Hiện. A few streets over, the votive paper market on Hàng Mã was bustling with shoppers during the last weeks of "ghost month," the seventh month of the lunar calendar. The acoustic dissonance I encountered between eerie quiet and lively activity underscored stark spatial disparities across the city, between pockets that resonated with the usual rhythms and rituals of daily urban life and those that seemed suspended in the stasis of an earlier pandemic time.

Sounding the "New Normal": Living with the Virus

By August 21, one month after the Đà Nẵng outbreak, the daily coronavirus infections in Vietnam had dropped to single digits, mostly imported cases. This decline came just as the country's total case count had reached one thousand. Two weeks later, on September 3, Vietnam recorded its last COVID-19 fatality for 2020, bringing the total number of deaths to 35. Community spread of the virus had again been swiftly contained, attributed to the government's immediate and coordinated action. People were eager to resume the path to post-pandemic normalcy, even as schools and international borders remained closed. Later that month, to stimulate the local economy, the Hoàn Kiếm District government launched Hàng Mã Street's annual Mid-Autumn Festival Market to celebrate Tết trung thu.[41] The occasion drew thousands of visitors, "overloading" (*quá tải*) and "paralyzing" (*tê liệt*) the streets (Phạm Duy 2020), Despite public health concerns, attendance had increased significantly compared with the previous two years.

On a warm September afternoon, after finishing a ghost town soundwalk, I headed over to Hàng Mã. There, I found throngs of people—local residents and domestic tourists from as far as Hồ Chí Minh City—reveling in the festive atmosphere of the Tết trung thu outdoor market. The environment was saturated with rich sensory stimuli: shops were adorned with exquisitely crafted lanterns, star lamps embellished with tinsel, colorful toys, lion costumes and paper masks.

Vendors lined the surrounding streets selling moon cakes and fast food. The area was cordoned off from traffic, and navigating through the dense crowds proved challenging and worrisome. Adherence to spatial distancing was all but impossible, and many people had forgone wearing protective masks in the heat. The deserted tranquility of the adjacent ghost town differed dramatically from Hàng Mã's loud and teeming streets, heightening my apprehension about social contact and the risk of airborne transmission.

The Mid-Autumn Festival signaled a pivotal shift in Vietnam's crisis response and its perception of pandemic time. Moving away from short-term, crisis-driven approaches to safeguarding life, the government began adopting a more sustainable, long-term model of pandemic management. This policy evolution involved abandoning wartime rhetoric of viral *conquest* to instead emphasize *cohabitation*, implying that the virus was here to stay. The "Thông điệp 5K," or 5K Message campaign, exemplified this shift, reshaping the content of state communications through image and sound.

Launched in early September by the Department of Preventative Medicine under the Ministry of Health, the Thông điệp 5K initiative reoriented governance strategies from a targeted focus on "fighting against" (*cuộc chiến chống*) the coronavirus to a holistic one of "safely living with" it (*chung sống cùng an toàn với*). A set of five precautionary measures for human-virus coexistence—*khẩu trang* (mask), *khử khuẩn* (disinfect), *khoảng cách* (distance), *không tụ tập* (no gathering), *khai báo y tế* (report illness)—embodied this new vision of balance among state oversight, public health, and social well-being. Framed within a context of a new normal (*bình thường mới*), this change in pandemic governance suggested resilient adaptation to the presence of the virus, rather than an expectation of its eradication.

During the mid-autumn festivities, the "5K" symbol was strategically displayed on signs throughout the market to remind visitors of the new social expectations and mindset toward the virus. Ironically, this municipal government-organized event, my first experience with densely packed crowds since the onset of COVID-19, seemed to contradict both the campaign's recommendations and its strategy for managing the next phase of the pandemic. Just outside the entrance gate, a police sound truck stood parked, unoccupied, its windows rolled down and mounted megaphone silent. This quieter presence, where the state no longer asserted its sonic authority, symbolized for me a transitional moment in the city's public health response. Were acoustic interventions, once central to crisis messaging, being phased out in favor of more subdued approaches to regulating life with the virus?

As fear of contagion subsided and emergency measures were lifted, the need for sonic modes of governance—and their broader applications in care work—diminished, even as Thông điệp 5K became part of my ward's public messaging. The sonic presence of the state waned accordingly, with loudspeaker broadcasts becoming shorter, less frequent, and less urgent, allowing other sounds to emerge. On October 1, 2020, the fifteenth day of the eighth month of the lunar calendar that

AUDIO CLIP 9. Mid-Autumn Festival soundscape: the lion dance unleashes ritual noise for communal pandemic protection. Source: author.

To hear this audio clip, scan the QR code with your mobile device or visit DOI: https://doi.org/10.1525/luminos.249.audio9

00:00:03 — 00:00:16

Audio clip 9

marked Tết trung thu, the sound of thunderous drumming outside my apartment sent me running to the street. There, I encountered a spectacle of lion dancers in elaborate costumes performing acrobatic feats to the delight of onlookers, accompanied by the clanging of cymbals and drums, right beneath the quieted loa phường (audio clip 9). The procession was led by a smiling Ông Địa, or God of Earth, whose kind, round face symbolized the moon. As the mythical lion creature romped about and lurched at the crowd, the music grew more frenzied.

While immersing myself in the totality of this unexpected sensory event, I began to see the performance as more than a source of cultural entertainment for the audience. Given the exceptional context in which the dance was unfolding—with much of the world still under the grip of the virus—I recognized its deeper symbolic meaning as a protection ritual, perhaps even a conduit for healing from COVID-19. Vietnamese custom holds that the noisier and livelier the performance, the greater the good fortune and the higher the likelihood of relief from hardship and suffering. This belief maintains that deafening clamor wards off lurking evil spirits—malevolent forces that, in 2020, included the coronavirus. As a ritual expression of collective care and renewal that brought the neighborhood together through ceremony, that year's lion dance unleashed a sonic barrage, an exceptionally auspicious omen for the times ahead.

Migrant Workers

The Philippine Singer (Ca Sĩ Philippines)

Travel restrictions and the closure of borders around the world left millions of people, especially migrant workers, stranded and unable to make their journeys home. Vietnam's suspension of international flights on March 25, 2020, shut people out of the country as much as it locked people in. That is, until repatriation flights could begin.[1] The press extensively covered the tremendous efforts made to repatriate Vietnamese migrants desperate to return to their families from distant places, including the case of 219 workers on infrastructure projects in Equatorial Guinea, of whom approximately 120 were reportedly infected with the virus (Hồng Vân 2020). These instances were portrayed as spectacular acts of state care under the motto, "no one left behind" (*không để ai bị bỏ lại phía sau*), despite being beset with bribery and favoritism that, three years later, would lead to an equally spectacular reshuffling of state and party leadership.[2] Less attention was given, however, to the plight of foreign migrant workers stranded in Vietnam, whose presence raised critical questions about the privilege that separated those who possessed the ability to move relatively freely, also across borders, from those whose mobility was severely restricted.

The pandemic violently disrupted the global mobility of people and goods, devastating migrant laborers and the remittance flows on which their families depended. Many lost jobs and were forced to return home at their own expense, while others found themselves trapped in precarious, overcrowded conditions lacking adequate social protections. Perhaps no situation better exemplified migrant stuckness than the hundreds of thousands of seafarers in the shipping industry, unable to disembark from their vessels at the start of the pandemic.[3] International media spotlighted the plight of Filipino crew members forced into a floating lockdown aboard dozens of cruise ships anchored in Manila Bay. Filipino nurses in the United States also found themselves stuck in the most precarious—and exploitative—frontline positions in health care, suffering disproportionately high COVID-related mortalities.[4]

Histories of imperialism and racial discrimination that deemed Filipino bodies "fit" for certain kinds of work not only shaped the industry of migrant health care providers (Choy 2003) but also that of overseas performing artists (OPAs), who deliver another form of affective labor (Balance 2016). Stephanie Ng (2005) has documented Filipino bands, often led by female vocalists, performing cover songs on the luxury hotel circuit across Asia as part of an export entertainment industry catering to middle- and upper-income countries. However, the migratory pathways of Filipino musicians extend beyond affluent destinations, as their performance skills are similarly in demand in emerging economies. This challenges the conventional view that labor migration flows exclusively from "developing" to "developed" countries. Overlooked South-South mobilities reveal alternative economic aspirations, exemplified by Filipino migrants attracted to Vietnam's expanding opportunities.[5]

Filipino musicians have long been fixtures on Vietnam's music scene, even before the country achieved lower-middle-income status a decade into the new millennium. Their presence spans from five-star resorts to live music venues like the now-defunct Seventeen Saloon in Hanoi, a popular cowboy-themed bar that hired Filipino rock bands to perform 1980s hits to mixed local and expat audiences. During the pandemic, some of these performers, unable to depart before border closures, continued their musical craft at upscale hotels. While scholars have examined the temporal relationship between crisis and aspiring migrants' sense of stuckness (see Hage 2009), this instance revealed how immobility created unexpected avenues for stranded entertainers to ease the pandemic anxieties of a listening public through sonic care. Such musical caretaking emerged alongside a new Vietnamese travel concept: the luxury staycation.

While much of the world confronted death and disease, life within Vietnam appeared to thrive as a COVID-free oasis of reassuring—though ultimately illusory—normalcy. Spatial distancing measures remained in place, albeit with inconsistent enforcement (see act 3). After the nationwide lockdown in April 2020, urban rhythms resumed their familiar cadence as the pandemic felt increasingly distant. The economy demonstrated uneven resilience, flourishing in some sectors, like domestic tourism, while faltering in others.

With international tourism halted, the hospitality industry slashed prices and reoriented marketing toward domestic travelers seeking discounted luxury experiences. By June, the hilltop villas at the Six Senses beachfront resort in Nha Trang were fully booked at US$2,000 per night—down from US$5,000, the Indian manager, the only other foreigner on a HanoiTourist day trip to Cúc Phương National Park, told me. Across the capital city, vacant five-star hotels drastically reduced rates to entice middle-class Vietnamese families to enjoy upscale staycations close to home, providing respite from both the heat and the risk of being quarantined while traveling. This strategy proved effective; my spouse and I booked premium venues like the legendary Metropole Hotel for just over $100, finding pools filled with Vietnamese children and restaurants hosting young couples enjoying the romantic ambiance. It was during these brief weekend escapes that I met the singer Cecilia, part of a three-person Filipino ensemble on Hanoi's entertainment circuit, and became a sonic witness to her art of musical caretaking.[6] Through her performances, I heard firsthand how stranded migrant musicians had become central to this reimagined tourism economy.

Staycations in Hanoi were contingent on the stuckness of Filipino performative labor. What constituted leisure for those indulging in deep discounts—myself included—was life in limbo for others. Cecilia and her bandmates had been in Vietnam for just over a year when the coronavirus forced the suspension of international flights, stranding them overseas. Her instinct had been to return home, she explained, but her work contract prevented her from doing so. With few social contacts in the city, she spent the three-week lockdown isolated in her small apartment. Afterward, her band returned to the hotel rotation, with a much-reduced performance schedule.[7]

Cecilia confided that she worried deeply about her family back in the Philippines. Pandemic disruptions had hindered her ability to send remittances home, where COVID-19 infections were surging. The forced separation during a global health crisis only compounded her distress. Unable to extend care directly to her loved ones, Cecilia spent her evenings caring for anonymous audience members by elevating their moods through the soothing sonic uplift of song.

Cecilia's affective labor of sonic caretaking reflected a gendered division of musical labor within the band, exposing how the pandemic amplified expectations placed on female performers. As the lead vocalist, she was responsible for interfacing with the crowd, while the male keyboardist and guitarist provided backup support. Spatially positioned closer to the listeners, she faced increased risk of virus exposure. As she delivered a striking rendition of Adele's "Hello" with impassioned vocals, she smiled warmly, making deliberate eye contact with each guest in their opening set. Between numbers, Cecilia invited the audience to submit requests for favorite songs, displaying quick sonic adaptability to emotional needs (Balance 2016, 79). Unlike other forms of affective labor requiring physical contact, her sonic care work remained effective at creating interpersonal resonance even while adhering to distancing protocols.

Cecilia's acoustic labor of managing affect had a temporal dimension: it was both present focused, attuned to the immediate performance, and future oriented. Her musical tending involved memorizing the preferences of returning guests to evoke feelings of pleasure and familiarity. This sonic memory work was exemplified one evening at a hotel lounge when, to our surprise, we found Cecilia and her band performing. As she noticed our arrival, Cecilia seamlessly transitioned into one of our requested songs from a previous set, warmly acknowledging our presence through a personalized music connection—an auditory gesture that transformed anonymous hotel guests into recognized listeners.

It was late, and the bar was mostly empty, though the adjoining restaurant's seafood buffet bustled with Vietnamese families, producing a contrasting acoustic register. I felt compelled to stay and order another drink, reciprocating the sonic relationship through attentive listening—lest they lose the venue—just as they had attended to our need for auditory sustenance. However, this was no equitable exchange of care. My conscious attempt at sonic reciprocity only underscored the asymmetry of our positions; while we could choose to extend our listening as a form of support, Cecilia and her band had to maintain their musical offering regardless of audience response. As listeners positioned relationally to the producers of sound and affect, we found ourselves enmeshed in the very power structures that precipitated their pandemic precarity, a vulnerability anchored in starkly uneven mobilities, both within urban acoustic spaces and across national boundaries.

Coda

Epi(demi)c Endings

How will we remember the critical first year of the pandemic, an unprecedented disaster that radically transformed the world and reshaped the fabric of society, while redefining relationships between humans, nonhumans, and the environment? What dominant narratives and memories will emerge to recount, or even resound, the pandemic as an epic social, epidemiological, and distinctly sonic event—a global catastrophe, felt and experienced locally?

Numerous scholarly and artistic projects, predominantly situated in Europe and North America, have been launched as part of digital humanities initiatives to archive the pandemic's immense impact on our sensory encounters. These projects have documented the deeply disorienting changes people experienced in their senses of smell, taste, sight, touch, and hearing—sensory reorientations in response to pandemic-induced isolation. A pervasive feeling of disconnection from daily rhythms of life, especially interpersonal contact, magnified these sensory distortions. For instance, pandemic archival projects document the disruption of tactile stimulation that reconfigured care practices and everyday haptic social customs (such as handshakes and hugs), the loss of smell and taste due to the virus's effect on olfactory and gustatory functions, changes to foodscapes (including the rise of so-called ghost kitchens), and changes to urban soundscapes, primarily recorded in cities in the Global North.[1] While these sensory shifts may to some extent seem universal, both their lived experience and the meanings attributed to them are historically and culturally specific. Moreover, these projects reflect scaled responses to vastly different pandemic conditions that unfolded unevenly across the globe, raising urgent questions about which stories—and sounds—are privileged in pandemic narratives.

Sonic Socialism—my modest autoethnographic contribution to this vast sensory memoryscape—frames the pandemic in auditory terms, centering on sound and its evolution during the first year of COVID-19. Spanning three acts that followed the timeline of the crisis while allowing narrative loops and echoes to mirror the temporal disjunctures of pandemic life, it examined the role of sounds in navigating the complexities of isolation in all its forms (quarantine, lockdown, and distancing), while also tracing changes to both the acoustic environment and the nature and contexts of listening. The book asked: If we listen carefully (and fully with care), what do sounds tell us about disaster preparedness and response in an epidemic of great historical—if not epic—significance? Sonic analysis goes beyond the mechanics of crisis management to illuminate how societies perceive, understand, and respond to catastrophes and how those choices reflect core values, particularly regarding who deserves protection and care, and how that care manifests in culturally distinctive ways. By engaging in attentive listening—in my case, "guest listening" with Western ears (Robinson 2020, 51)—this multisensory study reveals that crisis-informed actions emerged from complex cultural and ethical considerations rooted in historical experiences and patterns of sonic governance.[2]

My impulse, however, was not simply to document the upheaval of 2020 for future reference, as pandemic archival projects are wont to do, nor treat it merely as a repository of lived experiences. Instead, I sought to understand anthropologically the dynamic relationship between state and nonstate sounding practices— between government intervention and public response—while attending to those anarchival resonances often overlooked in conventional forms of documentation. One of my primary goals was to better grasp how a country with far fewer resources managed its public health crisis that first year more effectively than wealthier countries with advanced health care systems. I turned to sound-reproduction technologies not to trace direct causality, such as attributing outcomes to state-curated sounds, but to analyze how sonic interventions became vital tools for governing everyday life. These interventions, I argue, drew on long-standing traditions of social mobilization, practices I term "sonic socialism," rooted in the project of socialist nation building. Then as now, the Vietnamese state leveraged the persuasive power of sound to carefully calibrate a balance between social order, biosecurity, and public vulnerability. My analysis of these evolving sound-based strategies during the pandemic—and the spectrum of sonic dissent they provoked—underscores the fluid and adaptative nature of crisis governance without asserting the definitive success of any single method. This entanglement of sound, response, and re-sounding generated unexpected modulations in Hanoi's acoustic environment, reshaping the city's sensory landscape.

SONIC DISTORTIONS

Sounds compel listening, at times extractive "hungry" listening, as Robinson (2020) terms settler listening practices.[3] Embedded in historical relations of power, sounds operate through cultural, political, and technological forces that shape how and what we hear, forming what Sterne (2003) identifies as "audile techniques," the acquired practices that guide listening and regulate conduct. During the pandemic, these techniques cultivated listening sensibilities deemed appropriate for ensuring public health and national security. In consolidating power through sound, the Vietnamese state sought to uphold socialist values, particularly solidarity and productivity, while strengthening public trust in its institutions. However, this trust was severely shaken in 2022 by corruption scandals involving repatriation flights and the very COVID test kits that had been credited with keeping the country safe.

Deeply networked into urban infrastructures, sound facilitated the dissemination of information and provision of care to an imagined public audience, without defaulting to market-driven crisis solutions. Through emerging and residual media technologies—from catchy public health jingles to authoritative public service announcements—people were integrated into networks of care and accountability. These efforts aimed to enforce government containment policies and promote science-based hygiene thinking while combatting the spread of misinformation, especially superstitious beliefs. In the face of a deadly airborne virus, the stakes of auditory attention were high, from listening as compassionate care to listening as a source of fear, such as fear of the dry "COVID cough," which drove people to sonically surveil those suspected of exhibiting symptoms (Scott 2021).

Sound, however, was not always easily perceived or interpreted. The quality and clarity of sounds were just as crucial as their presence, affecting both communication and the comprehension of messages. Unstable re-recorded sounds required the constant recalibration of listening skills, especially for the foreign anthropologist who faced additional challenges in deciphering what became an auditory landscape of distortion. As Brian Larkin (2014, 1004) notes, the frequent use (and duplication) of sound-reproduction technologies at maximum capacity often leads to "tremendous distortion" of transduced sounds, a phenomenon that resonates with my own soundwork experience of distortion as both a material affordance and a regular sonic occurrence.

Distortion manifested at the nexus of material, mechanical, and acoustic effects in technologies overworked during the pandemic. Public address systems produced amplification-induced interference, while masks muffled voices, creating a barrier to clear communication. Microphones emitted high-pitched feedback and loudspeakers suffered from resonance distortion due to signal conversion at high amplitudes, further degrading audio quality. Weak signals led to radio static, while overdriven subwoofers struggled to produce clean bass frequencies. This tapestry of scrambled sounds showed material distortion to be inherent to the technologies themselves, shaping their ability to "act on" listeners to generate sensory effects and affects (Larkin 2004, 310). Distortion challenged my everyday interpretive capabilities, transforming the labor of listening into an intensely active and focused practice. It also metaphorically pointed to the incomprehensibility of the state behind its façade of seamless coherence.

FIGURE 28. Street art at the northern edge of West Lake features Việt rapper T.B.O alongside a residual sound-reproduction technology—the boombox—affirming both Vietnamese sonic modernity and state messages of collective responsibility in adhering to pandemic guidelines, 2020. Photo by the author.

Throughout the pandemic, people actively engaged with and reshaped state-produced sounds, in all their media forms and qualities, rather than passively receiving them. This dynamic urban soundscape both enforced conformity and enabled agency, often simultaneously. Official sounds meant to convey cohesion and reassurance were reinterpreted in creative ways that blurred the line between compliance and nonadherence. Unpredictable sounds, emerging at random moments, further reconfigured state sonic orders. These counter-practices made distinct sonic claims on the city, generating alternative acoustic traces as the basis of a sonic anarchive that departed from familiar tropes of war and heroism.[4] The dissonance created by overlapping sonic expressions exposed the futility of the state's attempts to control both the acoustic environment and narratives about the pandemic.

Consider, for instance, the lion dance from act 3, a traditional Mid-Autumn Festival celebration transformed into a protection ritual against COVID-19, with its thunderous cacophony meant to ward off the virus through auditory rather than biomedical interventions. As the dancers' noise reverberated across the lake, asserting sonic dominance over public space, their performance revealed tensions between folk traditions and official scientific discourse, while also serving as a powerful symbol of resilience, recovery, and a hopeful return to normalcy.

Not all non-state-produced sounds contravened public health directives. Some aligned with government policies and socialist values by echoing state messages about COVID-19. Vietnamese rap (*rap Việt*), considered an "underground" music genre (*dòng nhạc underground*), offers a compelling example. During a summer 2020 bike ride around West Lake, I came across a colorful street art mural depicting the Việt rapper T.B.O (figure 28). Wearing Adidas gear with a blue bucket hat and a white protective mask, he stands against a cityscape backdrop, juxtaposing socialist collective housing (*khu tập thể*) with modern high-rise buildings. The English block-letter text, "stay home, wash your hands, be safe," reinforced state biomedical messaging, while the resounding boombox at his feet, a street technology whose productive distortion echoes "sonic Afro-modernity" (Weheliye 2005), illustrated how Việt rappers interwove top-down sonic discipline with grassroots cultural expression. In this way, countercultural forms amplified official health guidelines (such as handwashing and mask-wearing) while preserving a distinctive sonic character, shaped by a global hip hop aesthetic.

A COUNTRY OF PLANNING

The Mid-Autumn Festival lion dance proved an effective ritual for safeguarding the country and its economy through 2020. The release of annual statistics at the end of the year garnered accolades for Vietnam from across Asia. COVID-19 was Vietnam's "breakout moment"; it was the "country that got it right" (Hoang 2021). With only thirty-five deaths and just under 1,500 cases by the end of 2020, it possessed one of the lowest figures on the globe. Vietnam's COVID numbers made the country "the envy of the world," also because of its "stable economy." Where other countries floundered, Vietnam flourished as the "sole economic winner in Southeast Asia," outperforming its ASEAN peers with one of the highest GDP growth rates in the world at 2.9 percent. It was poised to become, as predicted in a media headline on November 19, the "single economic success story in the coronavirus era" (Nakano and Onishi 2020).[5]

On that same day in the United States, the media reported that consumers were once again panic-buying toilet paper and other items amid rumors of another lockdown. This surge in demand prompted stores to reintroduce rationing of basic supplies due to concerns about disruptions in the supply chain. Not long after, news broke that the death toll from COVID-19 was approaching three hundred thousand Americans, an alarming increase from two hundred fifty thousand only three weeks earlier, leading the world in confirmed deaths. This was five times as many deaths as Americans killed in Vietnam, the *Washington Post* estimated, underscoring how the war continued to serve as a benchmark for measuring the magnitude of human loss well into the pandemic. Eleven months after the first detected case in the United States, "*There [was] still no national plan*," the article noted disapprovingly (Fisher, Wilson and Hernández 2020, emphasis added). In contrast, Vietnam, marked by a history of millions of war-related casualties, had swiftly formed a national COVID-19 steering committee, drawing on its strong tradition of planning (*quy hoạch, kế hoạch*). How might we understand these stark disparities in pandemic experiences, responses, and outcomes during 2020, between an affluent nation suffering massive loss of life and a country with con-strained economic capacity achieving remarkable preservation of life?

Vietnam's early containment efforts challenge a common assumption that wealthier democracies, with their presumably higher levels of public trust, are better equipped to handle such crises. Vietnam's case instead affirms how strong state capacity, when combined with coordinated government action and broad public and institutional support, proves crucial in managing declared emergencies. While Western critics often characterize Vietnam as illiberal and nondemocratic, its decisive pandemic response demonstrated considerable trust and cooperation, enabled by centralized, hierarchical planning—a striking counterpoint to the fragmented responses seen elsewhere.

I introduced the concept of "disaster socialism" to account for Vietnam's his-torically informed approach to crisis management, which differs significantly from that of countries like the United States. This framework, grounded in prin-ciples of socialist paternalism (see Verdery 1996, 24–25), positions the state as a protective force steering society toward the common good. It clearly and sharply diverges from "disaster capitalism," a phenomenon identified by Naomi Klein and others, where market-driven actors systematically capitalize on mass catastrophe by prioritizing corporate interests over public welfare. Under disaster capitalism, a decentralized state allows corporations to appropriate common resources during moments of collective trauma rather than acting to benefit society as a whole.[6]

Disaster socialism is neither exclusive to Vietnam nor limited to the pandemic; socialist and democratic socialist countries with robust public investments have had a long history of proactive responses to disaster threats. Rather than reacting in the aftermath, these nations emphasize preparedness, guided by a governing logic that invokes the protection of people above the preservation of property as a moral imperative. This politics of life involves mobilizing populations in antici-pation of catastrophic events, such as evacuating millions of people when major floods are forecast.[7] Such anticipatory governance demands a high level of central-ized coordination, social cohesion, and comprehensive planning across multiple constituencies to mitigate risks to national well-being. By embracing coopera-tion over competition, disaster socialism advocates collective action and shared resources rather than private sector interventions.

While standard governance techniques employed during declared emergen-cies are generally designed to "sustain order and preserve life" (Lakoff 2017, 20),

Vietnam's disaster socialist approach reflects deeper forms of social and political organization that evolved from shared experiences of crisis, especially in wartime, which fostered networks of solidarity and mutual support that were echoed during the pandemic. Historic events like the victory over French colonialism at Điện Biên Phủ in 1954 exemplify how strategic planning and collective action emerged under extreme adversity, sometimes at tremendous human cost. Despite infrastructural constraints, Vietnam's ability to mobilize human capital in times of crisis—and not always voluntarily—highlights the social and historical dynamics that underpin its integrated emergency planning today, including those sonic practices that remain integral to its disaster response.

SOUND IN THE TIME OF DELTA

This autoethnography shifts the focus of disaster planning and preparedness scholarship from the visual (such as mapping and media representation) to the auditory, emphasizing the significance of sound in the provision and regulation of care. The acoustic qualities of this governance, anchored in specific temporal and spatial contexts, dynamically responded to localized emergency imperatives. A deep examination of these sonic facets yields new insights into the fluid sensory dimensions of crisis management under conditions of disaster socialism.

As socialist systems evolved, their sonic responses to crisis changed accordingly. The analog practices of sonic governance during the war for liberation resonated differently from the digitized sonic socialism of the pandemic. These auditory shifts indexed broader transformation in frameworks of disaster socialism, as the socialist state transitioned from a command economy to a model of market socialism that utilized regulated market mechanisms to achieve its ostensible goals of equality and social well-being.[8] Take redistribution, one of the fundamental paternalist principles of state socialism, as Verdery notes (1996, 24), that continues to hold significant weight in Vietnamese society.[9] Under a market socialist model, disaster socialism sought to align industries, particularly those with government ties, with expectations of resource reallocation in support of vulnerable social groups. In the business sector, this manifested through initiatives like mask or rice ATMs and poster competitions, highlighting the moral virtues of social entrepreneurialism and its associated caretaking of communities. In the cultural sector, the state leveraged popular culture in its multimedia health agenda, collaborating with well-known musicians to produce public service videos set to popular songs, blending commercial interests with social welfare objectives.

The promise of state or ritual protection proved impossible to sustain during the second year of the pandemic, however. The protective powers of the lion dance against evil spirits—and, by extension, the coronavirus—were no match for the Delta variant, which reversed the positive trends of 2020 and plunged the country into a downward spiral starting in late April 2021, one year after the first national lockdown. This turn of events highlighted both the limitations and flexibility of disaster socialism. As a resurgent public health crisis gripped the nation, the state once again adapted its approach, reverting to its "Zero-COVID" policy and use of martial rhetoric. The public loudspeaker in my neighborhood would proceed to broadcast a familiar soundbite: "Fighting the pandemic is like fighting the enemy."

The stark contrast between Vietnam's initial success and its later struggles is reflected in the numbers: Through the end of 2020, the country had maintained an exceptionally low death toll of just thirty-five. That figure remained steady until May 15, 2021, when the country recorded its thirty-sixth fatality. However, as pandemic fatigue set in and the more contagious Delta virus took hold, the situation changed dramatically, with nonessential businesses closing again on May 25. Deaths increased gradually, surpassing one hundred cumulatively on July 7, before surging into the hundreds daily in August. The government's response—a protracted national lockdown through September, including a military-enforced stay-at-home order in Hồ Chí Minh City, where the majority of cases and deaths occurred—intensified efforts to curb the more virulent Delta strain. By October 2021, when I reentered the country as the "family member of an expert" and underwent a ten-day quarantine at a state-approved hotel facility, Vietnam's status as the region's enviable success story had been thoroughly upended, with statistics indicating an estimated twenty thousand deaths and more than eight hundred thousand cases of infection.[10] The Ministry of Health later acknowledged that reported numbers, especially in Hồ Chí Minh City, the epicenter of the outbreak, were likely to be much higher.

The deadly Delta wave offers important insights into disaster socialism in action, including its acoustic interventions as a means of governance. Sounds were ramped up, while new forms of care emerged, alongside accusations of carelessness. Large-scale, nationwide mobilizations, which continued to be coordinated through sound-based media, likely saved lives over the long term.[11] For example, the mass vaccination program, despite its slow initial rollout that contributed to higher death rates early on, gained momentum rapidly. Within five months, 90 percent of the population was fully vaccinated, with peak days seeing over 2.1 million doses administered with COVAX support.[12] During the final weeks of national lockdown, Hồ Chí Minh City maintained an especially strict shelter-in-place order enforced by security forces, which necessitated the distribution of food to over ten million people, coordinated through wards and neighborhood leaders with assistance from the military. Harsh measures led to more rigorous policing, resulting in incidents of power abuse and reports of corruption.[13]

Despite these allegations, public trust reportedly remained high, with 84 percent of citizens surveyed by UNDP and the Mekong Development Research Institute in 2021 expressing approval of the government's response to the Delta surge. This marked, however, a substantial decrease from 97 percent the previous year, indicating the pandemic's growing negative impact on families.[14] A striking 93 percent supported strict containment measures (with 70 percent expressing strong support), despite one-third of respondents reporting difficulties accessing food or emergency relief in high-impact areas during the extended lockdown. This reveals significant logistical challenges that the state faced in ensuring food security and adequate assistance, especially in poorer peripheral areas, where mutual aid networks emerged to address unmet needs, operating independently of government oversight. At the same time, a surprising 96 percent of those surveyed approved of the involvement of the police and military in pandemic prevention efforts, demonstrating public alignment with the whole-of-society strategy as security forces filled gaps left by an overwhelmed state, particularly at the level of local government.[15] Overall, the survey indicated a strong preference for saving lives over economic recovery (83 percent), embodying a core principle of disaster socialism outlined in this book, namely, that collective well-being should override economic considerations during crises. While this model of governance may have aspired to prioritize life, it did not always achieve this goal successfully or equitably. The high Delta death rate revealed its shortcomings, as did the uneven distribution of vaccines, with Chinese-made Sinopharm doses, rejected as inferior by urban residents who favored Astra or Moderna, designated for use in the countryside.

The evolving sounds of the Delta wave of COVID-19 reflected this complex and unraveling terrain, highlighting the contrast between the messy reality of actually existing (disaster) socialism and its hopeful realization. At the onset of this tumultuous period, the National Steering Committee called for intensified "multimedia public communications efforts using loudspeakers and radio in communes, wards and collective housing, as well as through the internet and mobile platforms for the dissemination of information," my ward broadcasted in May 2021.[16] Two years later (in 2023), while lecturing on medical listening at the University of Social Sciences and Humanities at Vietnam National University, Hồ Chí Minh City, I asked my students, "Which dominant sounds informed your experience of the Delta-stricken city?" Their responses offered a vivid audio portrait of urban crisis: More than 30 percent reported hearing ambulance sirens (*tiếng xe cấp cứu*), a sound absent in 2020, revealing the state's failure to guarantee the public safety of the population. Loudspeakers remained a staple of daily sonic listening in their wards, my students recalled, though their content shifted to raise awareness about mass testing and vaccine appointments at local health stations, similar to the messages in my neighborhood. Students also noted the presence of sounds intermingled with smells, from army trucks mass spraying disinfectant in city streets to soldiers and local authorities delivering food and essential goods to their residences. Their recollections powerfully evoked shifts in multisensory experiences of isolation across pandemic time.

Between online classes while confined to their homes, students spent a great deal of time listening to music, including Vietnamese hip hop. Much like rappers in Cuba who emerged as part of the revolutionary vanguard, Vietnamese rappers become participants in the pandemic vanguard. During this crisis, the Vietnamese state "harness[ed] the oppositional potential of underground rappers to maintain its hegemony" (Fernandes 2006, 127). While shaped by capitalist consumerism, the rap Việt scene also embodied elements of disaster socialism, including the moral expectation of social obligation. As noted above, these rappers aligned their underground movement with government directives, suggesting a form of state collaboration or even cooptation.

Throughout this book, I have detailed how the government used sounds to cultivate scientific rationality. So too did Vietnamese hip hop, as evidenced by the street art featuring T.B.O. The Việt rap group Space Speaker also embraced music as a force for social good at the height of the Delta lockdown in August 2021. As quarantine fatigue set in, the band launched a music challenge called #MuzikDapDich (Muzik Dập Dịch) or "music to stamp out the pandemic." This initiative served as a "musical vaccine" (*vắc xin âm nhạc*) to boost morale, combat food insecurity, and encourage adherence to health guidelines. The "turning music into rice" (*góp nhạc thành gạo*) campaign generated over twelve thousand kilograms of rice donations for the Fatherland Front in Hồ Chí Minh City, filling a gap left by the absence of adequate state protections, as the Delta variant overwhelmed existing systems.[17] This artistic mobilization reflected values emblematic of state socialism, such as resource redistribution and collective responsibility (figure 29). Rappers like Wowy, performing in frontline workers' hazmat suits, urged their fans to prioritize public health over personal desires by sheltering in place—or face discipline, another rapper warned. "To stay at home in peace," a collaborating artist rapped on YouTube, "one needs to hear the sound of music" (*Để ở yên trong nhà cần có tiếng ca*), harmonic rhythms offering emotional solace in contrast to the mechanical monotony and distortion of state vocalizations—forms of sonic governance explored throughout this book.[18]

Similar to state sonic campaigns, #MuzikDapDich rappers wove solidarity into their musical repertoire, revealing how the state's framework of crisis management permeated even countercultural movements. Youth, like my students, who typically tuned out intrusive pandemic messaging found themselves powerfully interpellated by this music, pulled into its acoustic realm, as pandemic-era rap imbued the soundscape of disaster socialism with new rhythms of urgency and care.

. . .

This sensory autoethnography has offered one perspective on Vietnamese crisis governance, which mobilized sound and sound-reproduction technologies to reach and integrate vulnerable communities into broader collectivities of care. While there are various ways to narrate that first year of an epic global pandemic, my analytical focus on sound provides a framework for understanding the sensorial and technological effects of Vietnam's preparedness and public health response

FIGURE 29. COVID-19 prevention poster in Hanoi emphasizes the protection of family and society, embodying Vietnam's relational approach to care-centered public health, 2020. Photo by the author.

through what I term sonic practices of care. These practices cultivated strong social cohesion initially, contrasting with the acrimony that came to define other parts in the world where trust in government, science, and fellow citizens had eroded. But solidarity was not absolute. Disaster resilience research has shown that while crises can swiftly unite people around a common purpose, that sense of unity can quickly dissolve when expressions of self-interest reemerge (Oliver-Smith 1999).

In the introduction, I raised the question of how Vietnam's response to national emergencies might offer valuable lessons for managing future crises. What distinguished Vietnam's sonic socialism was its resourceful adaptability, which played a key role in navigating the threat of catastrophe. This included its capacity to incorporate both top-down government directives and bottom-up creative endeavors, along with the mediations that unfolded in between. From loudspeaker broadcasts to underground hip hop, state messaging to street art, red music to V-pop, pandemic soundscapes embodied resilient socialist values that found new resonance and significance, embraced and contested across generations of urban audiences. This dual nature of socialist governance—balancing participation with strict control and oversight—has endured through crises by creating (and regulating) spaces where official and public voices that affirm state agendas amplify one another in collective efforts to safeguard human and nonhuman flourishing. I proposed the concept of sonic socialism not as a utopian imaginary or as a solution to the immense disparities observed between the achievements of 2020 and the failures of 2021 but as a starting point for rethinking how we assess the value and fragility of humanity—and the role of sound in upholding and communicating those values—in support of a shared commitment to caring for all forms of planetary life.

PREFACE

1. The video is based on the 2017 pop song "Ghen" ("Jealousy"), performed by Min and Erik and written by Khắc Hưng. As part of the state's crisis communication response to COVID-19, Khắc Hưng rewrote the lyrics to fit the context. By June 2020, the original video with English subtitles had 87.5 million views (www.youtube.com/watch?v=BtulL3oArQw). The subsequent English-language version released on April 9 featured sign language interpretation to include the Deaf community in the acoustic spaces of public health messaging (www.youtube.com/watch?v=wGoodWEtV8c).

2. By contrast, the United States had recorded more than 150 thousand deaths on that same day, an alarming one-fifth of the global toll, which stood at an estimated 750 thousand deaths as of July 31, 2020, as reported by the World Health Organization (WHO). See https://ourworldindata.org/covid-deaths.

3. Tên của em ấy Corona [Their name is Corona]
   Em từ đâu? Quê của em ở Vũ Hán [Where are they from? Their homeland is Wuhan.]

The English version of "Ghen Cô Vy" omitted this reference to Wuhan, likely due to concerns that associating the virus with a specific place had fueled anti-Chinese sentiments globally, including in Vietnam. Despite this change, the revised lyrics—"Out of the blue, it's coming for us"—still employed precisely the kind of military metaphors that Susan Sontag (1978, 65–66, 74) identified, where the spread of disease is portrayed as hostile invasion or cross-border infiltration. It is worth noting that at the time of "Ghen Cô Vy's" production, the term "Wuhan virus" dominated Western media, racializing the disease and framing foreignness as a threat. The video's adaptation for an international audience with more neutral language suggests Vietnam's sensitivity to these harmful connotations.

4. In the English-language version, the emphasis shifts from Vietnamese nationalism to internationalism. The national flag is replaced by an image of the globe, and

above it, hands of different skin tones form a heart, symbolizing global unity against COVID-19.

5. Sounding, as Henriques maintains in his book *Sonic Bodies*, is both reciprocal and relational, encompassing "everything, everyone and all the activities" (2011, xxix)—and, I would add, the materialities—that are invested in the active making and listening to sound.

6. The speech by Director-General Dr. Tedros Adhanom Ghebreyesus can be found at www.who.int/director-general/speeches/detail/who-director-general-s-opening-remarks -at-the-media-briefing-on-covid-19---11-march-2020.

7. Reassuring communications from campus administration downplayed the potential severity of the situation, contributing to this lack of urgency. A March 5 email from the chancellor, only five days before the directive to transition to online, stated, "UCR has been closely monitoring the status of the Coronavirus Disease (COVID-19). It remains impossible to determine if, when, or to what extent our campus might be impacted. However, should the virus continue to spread, it is likely that some faculty, staff and students will be affected and will have to remain quarantined or isolated for a period of time. Although it seems unlikely at this time, it is even possible for disruptions as significant as closing all or part of the campus on short notice for two weeks or more." Just two days later, Riverside County confirmed its first case of the virus. On March 16, the campus was abruptly closed. It would be two years before I could set foot on its grounds again, a duration that seemed unimaginable at the time.

8. Historical data on the total numbers of cases reported in the United States beginning January 22, 2020, can be found on the CDC website, www.cdc.gov/coronavirus/2019-ncov /cases-updates/previouscases.html.

9. Such global health declarations are politically charged, Andrew Lakoff (2017, 3) argues, raising critical questions about disease prioritization and the ethical obligations wealthy nations have toward poorer countries.

10. In an interview with the *New York Times*, Dr. Rajiv J. Shah, a health economist, physician, and current president of the Rockefeller Foundation, commented that ubiquitous testing could have averted the catastrophe: "I think history will record that the failure to get testing right in America . . . turned what could have been a manageable health crisis into a global pandemic that has wiped out U.S. $28 trillion of economic value" (Gelles 2021). Crisis declarations thus framed the pandemic's impact as multifaceted, encompassing the loss of both human life and economic stability.

PANDEMIC FIGURE 1: PATIENT 17

1. While I adopt pseudonyms for the other pandemic figures, here I follow the media's practice of using Patient 17's initials to highlight the particularity of this case.

2. Because of the significance of this superspreader event and the dangers it posed for air travel, Vietnam Airlines flight 54 became the subject of a Centers for Disease Control and Prevention (CDC) study of the risks of airborne transmission on commercial jets (Nguyen et al. 2020).

3. See also Joey S. Kim's (2020) piece "Orientalism in the Age of COVID-19," published one month later in the *Los Angeles Review of Books*.

4. Reflecting Sontag's (1978) observations in *Illness as Metaphor*, Sonia Shah (2020) noted about the coronavirus, "The ideas framing the outbreak emanate from an old

paradigm about contagion . . . [as] a problem of microbial invasion, a foreign incursion into domestic bodies to be repelled by military might."

5. While the official policy at the time was to withhold names and refer to cases solely by their assigned numbers, details such as place of residence and travel itineraries were publicly disclosed to facilitate contact tracing. However, this disease-control strategy changed following the social media backlash against Patient 17. To preserve some degree of anonymity, exact addresses were omitted, but residential areas or buildings were included in the interest of public safety. Nonetheless, detailed movement histories continued to be provided to the public to identify potential sites of cross-contamination.

6. For an anonymous response published in the English-language *Việt Nam News* that defended the government's handling of the outbreak, see "The Truth about COVID-19 Patient 17 in Việt Nam" (2020). In the *New Yorker* piece, NTN's sister chalked up the social media attacks to class envy and likened their treatment to racism, suggesting that, "If this was Paris Hilton, there would not be so much fuss."

7. For the WHO's guide to preventing social stigma associated with COVID-19, see www.who.int/docs/default-source/coronaviruse/covid19-stigma-guide.pdf.

8. For a video of Trúc Bạch Ward leaders on *Morning Coffee* dancing to "Ghen Cô Vy," see Phương Nhung (2020).

INTRODUCTION

1. By "soundscape," I mean the totality of sonic modalities in a given location, encompassing not only the sounds themselves but also the perception and interpretation of those sounds, which mediate relations between humans and nonhumans within a shared acoustic environment. Soundscapes—always in the plural—are never static; they are continually evolving as I show in this book, particularly in the context of pandemic time. Readers interested in the origins and popularization of the term should begin with R. Murray Schafer's (1994) foundational work on acoustic ecology, *The Soundscape: Our Sonic Environment and the Tuning of the World* (originally published in 1977 under the title *The Tuning of the World*). While debates continue about the usefulness and limitations of this now widely applied concept (see, for example, chapter 11 in Ingold [2011]), Schafer's work also notably explores "sound imperialism." By this, he means the weaponization of sound to dominate people and territory through "subjugation by Noise," where those wielding sufficient "sound power [can] create a large acoustic profile" (1994, 77). Following this logic, entities like the Vietnamese state can be seen as "imperialistic" in their tactical use of sound amplification technologies, originally developed by imperial powers, to establish dominance through sonic presence (1994, 91).

2. Innovative anthropological work on the governance of urban sounds can be found in Lippman (2021), Hsieh (2021), Peterson (2021), and Cardoso (2018), among others.

3. Various geopolitical, feminist, cultural, and genetic theories attempted to explain why certain countries, many in the Asia-Pacific region, fared better than others during the pandemic, some relying on essentialist logics. These ranged from the impact of female leadership (the women-as-better-listeners-and-caretakers argument) to the influence of geography on isolation and border control, particularly mountainous or archipelagic

geographies. Alleged genetic differences were suggested as influencing immunity across populations. Cultural factors like "Confucian values" presumed to cultivate respect for authority and collective thinking were proposed, as well as Indigenous perspectives on human/nonhuman collectivities promoting group safety. Some have speculated on the role of political systems, particularly how authoritarianism might have enforced stricter containment policies. Others emphasized historical perspectives on disaster resilience. New Zealand exemplified multiple proposed factors: female leadership, island geography, and Indigenous Māori ontologies of Manaakitanga that inspire acts of mutual care, also for the planet. These speculations and their application to Vietnam are explored in subsequent chapters with the aim to provide a more complex understanding of the country's ability to limit viral spread during the pandemic's first year.

4. For context, the three-year cumulative infection figures for Japan and South Korea were triple those for Vietnam. Vietnam's numbers were more comparable to Taiwan's and Australia's reported infections. Notably, Japan also recorded the highest number of deaths among these countries.

5. Not all orchestrated approaches to disease control have produced outcomes as favorable as those seen during the SARS and avian flu responses. For instance, health experts identified the Vietnamese government's failure to implement more effective disease containment measures following cholera outbreaks in 2007 (Lincoln 2014, 343).

6. The roots of Vietnam's public health infrastructure can be traced back to French colonial medicine and the coercive measures it imposed to manage contagious bodies (Monnais 2006). This historical context aligns with broader colonial legacies of disease control as a form of domination, such as "governing through contagion" in British colonial and postcolonial Singapore (Chua and Lee 2021), and the role of "tropical medicine" in the racialization of disease during American colonization of the Philippines (Anderson 2006).

7. On the condition of occupying the position of "Double Other," see Prats (2002). On the cultural politics of student radicalism during the Vietnamese revolution, see Tai (1992).

8. I borrow the phrase "Cold War Orientalism" from Klein (2003).

9. On the Vietnamese government's response to what it calls the "catfish campaign against Vietnamese fish" in a refusal to use the US term "catfish war," see https://vietnamembassy-usa.org/relations/catfish-campaign-against-vietnamese-fish. US media have consistently employed militarized metaphors to describe this ongoing dispute as a "trade war," since it first erupted in 2001. For instance, an early report opened with, "This time, the Vietnamese have invaded the United States, with catfish, and a bitter war has broken out for access to America's frying pans" (Mydans 2002).

10. According to a study conducted through July 1, 2020, the ten worst-affected countries with the highest mortality rates (measured by deaths per million inhabitants) were all ranked among the top twenty countries with the highest global health security (GSH) index scores (Haider et al. 2020). The GHS Index assesses the infrastructural capacity of countries to respond to infectious disease outbreaks.

11. On the relationship between American exceptionalism and white anxiety in antebellum historiography, see Glickstein (2002). Note that this "COVID reversal" would itself be reversed one year later. Vietnam's inability to secure access to vaccine doses from well-supplied countries would reestablish a North-South power asymmetry as the country became dependent on COVAX donations to control Delta outbreaks in 2021.

12. See, for example, David Welna's (2020) piece "Coronavirus Has Now Killed More Americans than Vietnam War." For a thoughtful commentary on the tendency to equate US body counts in Vietnam with those of the pandemic, see Bui (2020).

13. That the pandemic presented an opportunity to reduce a country with a long and vibrant history of achievements to a painful chapter of protracted violence was evident in the surge of interest in the martial metaphors utilized in Vietnam, as if such language was unique to that context. This narrow focus overlooked how militarized rhetoric in the United States and elsewhere, including references to the Vietnam War as both a benchmark and a cautionary lesson, has escaped critical scrutiny. For an example of combat terminology in the context of US public health policy failures, see *Lessons from the Covid War* (Covid Crisis Group 2023). For an example of "Vietnam" used as a global political metaphor, see Herrero and Ramzy's (2019) article "Maduro Sounds Conciliatory but Warns: US Intervention Would Be Worse than Vietnam." To discourage such associations and take control of a narrative framed through American experiences, the Vietnamese government coined the phrase "Vietnam: A Country and Not a War" in 1995, following the opening of its first embassy in Washington, DC (Schwenkel 2009a).

14. The report luridly described black cats as "boiled, skinned and cooked before being turned into paste" and sold as medicine against the coronavirus (Sun Media 2020). This accusation of wanton animal torture is steeped in Orientalist tropes of Asian brutality and disregard for humanity, echoing the racist Vietnam War–era mindset that "life is cheap in the Orient."

15. Modern disasters, especially warfare, also open up possibilities for reimagining alternative futures in the effort to rebuild social, economic, and material worlds after catastrophe, as I have demonstrated in the case of Vietnam (Schwenkel 2020).

16. Feminist theorists have rightly pointed out the complexities surrounding the notion of "consent," calling into question declarations of action as "consensual." Carla Jones (2020), for instance, proposes the term *consentir* as a mode of consent achieved through co-feeling and comity, rather than through potential coercion.

17. See Lacey (2013, 30) on the relationship between print media and visual culture that led to the "de-auralization" of the public sphere.

18. Nonbinary does not infer the absence of hierarchy. However, even though listening publics are never free from power dynamics, power does not determine or predict behavior but instead creates opportunities for agency to navigate the fluid boundaries between refusal and accommodation. Lacey's (2013, 165) emphasis on the active "response-ability" of listening draws attention to the intersubjective encounter or resonance between apparatus and audience, speaker and listener, rather than a one-way transmission of information that consolidates state authority.

19. The media technologies discussed in this book were intended to manage not only practices of care but also public affects of fear. Fear mediated relationships between citizens and the state, as well as between humans and nonhumans (like the virus), as people navigated pandemic uncertainties. My interlocutors expressed fear (*sợ*) in relation to perceived risks, which underpinned their logic of working with rather than against the state. These fears included fear of infection (and mandatory care in isolation units), fear of death (of self and others, particularly the elderly), fear of hospitals (places where people go not to heal but to die), fear of economic insecurity, fear of unknowingly transmitting the virus, especially to grandparents in multigenerational households, and fear of social shaming if any of the above were to occur.

20. See, for example, the Wenner Gren forum on the future of anthropological research, at https://wennergren.org/forum/the-future-of-anthropological-research-ethics-questions-and-methods-in-the-age-of-covid-19/.

21. This practice, however, raises critical questions about power dynamics underpinning authorship and knowledge production.

22. On the intersections—and tensions—between multimodal and sensory scholarship, see Pink (2011).

23. While "being there" afforded me certain advantages, it does not imply that my account of the pandemic is inherently more authentic or objective than those produced from afar. Nor can I speak for Vietnamese experiences. I present, quite simply, a situated narrative of the pandemic from a particular embodied perspective, grounded in my own sensory positionality and presence in the country as a recipient of the monthly "humanitarian visa extension" offered to foreigners by the Vietnamese government. Nonetheless, multisensory immersion does allow for a unique experiential standpoint on social phenomena. As I craft this introduction from my Trúc Bạch apartment, I am engulfed by the ambient sounds of pagoda chimes, honking motorbikes, clamorous construction, calls of broom sellers, and a blaring megaphone at the Vietnam-Cuba primary school. This sonic cacophony, which I came to relish and appreciate even more following the silent months of the pandemic, shapes how I write, hear, perceive, and engage with the city through my own Western feminist sensibilities.

24. On the sensory hierarchization of human races, tethered to stages of social evolution that reproduced ideologies of racial difference, see Classen (1997, 405).

25. This is not to deny the role of visual modalities during the pandemic. I recognize that sensory registers "cooperate so closely . . . that their respective contributions are impossible to tease apart" (Ingold 2011, 136). However, although the senses may work in tandem, they do not always operate equally or consistently with one another.

26. There is a growing body of literature on the anthropology of sound and sonic histories that emphasize listening as epistemology and research methodology, including Marina Peterson's 2021 ethnography *Atmospheric Noise: The Indefinite Urbanism of Los Angeles*, Ana María Ochoa Gautier's 2014 book *Aurality: Listening and Knowledge in Nineteenth-Century Columbia*, and Stefan Helmreich's 2016 collection of essays *Sounding the Limits of Life*.

27. For an exception, see Robert (2020), an anthropologist in Hồ Chí Minh City during the pandemic.

28. Starting in the 1870s, hygiene was taught orally in Indochinese classrooms, with written materials and textbooks only introduced after 1906 (Monnais 2006, 42n5). The fact that Vũ Trọng Phụng was writing about late 1930s Hanoi demonstrates the continuation of oral pedagogies to promote hygiene and sexual health.

29. Vũ Trọng Phụng (2011, 54, 158n8) recounts the government's "Ballad of Eros," which sex workers confined to the medical dispensary were required to memorize by ear and recite as a condition for their release:

> If you meet a client,
> Be pleased to let him pick the flower.
> But before everything, you must wash things clean.
> As you do with yourself, tell your client he must follow;

> White soap, clear water;
> There's nothing to worry about washing together.

30. For an example of party discourse during the pandemic that stressed the need for society to "serve people" (*phục vụ con người*) rather than "only pursue profits" (*chỉ theo đuổi lợi nhuận*) and "harm human dignity" (*làm phương hại đến phẩm giá con người*), see Trần (2021).

31. On the role of the media in forging narratives of "disaster nationalism" in response to climate-related catastrophes in the Philippines, see Guyton (2022).

32. I use "social types" and "pandemic figures" in the spirit of the edited volume *Figures of Southeast Asian Modernity* (to which I was a contributor), edited by Joshua Barker, Erik Harms, and Johan Lindquist (2014).

PANDEMIC FIGURE 2: THE SECURITY GUARD

1. All names are pseudonyms to ensure anonymity.

2. Here I extend Jonathan Sterne's concept of "techniques of listening," which he defines as a set of culturally acquired practices and relational orientations toward auditory perception (2003, 90).

ACT 1: QUARANTINE

1. For the controversial application of Agamben's (2021) own insights to critique state responses to the pandemic and what he viewed as the weaponization of the crisis, see his collection of blog posts in *Where Are We Now?* For an analysis of his disputed comparison between pandemic lockdowns and Nazi Germany, and the subsequent academic backlash it sparked, see Kotsko (2022).

2. Zalo is the leading Vietnamese social media and messaging platform, with tens of millions of users, and is considered the country's first unicorn.

3. A fellow anthropologist of Vietnam once remarked that "no one can be sick or hurting in Vietnam without some local organization (Red Cross, Women's Union, Veterans [Association] . . . etc. etc.) knowing about it, and being charged to offer such help as they were able to." Email communication, April 28, 2020. In other words, what appears to be supervision can also provide the possibility of care.

4. The tweet from March 22, 2020, read, "Village VN police texted us to ask about our health after our flights back to Hanoi. We sent a picture of us with masks on watching tv. He sent a meme back with a pumping heart and asked if we were really wearing masks inside. :-)"; https://x.com/mallonray/status/1241717065649491971. In a May 13, 2020, group email, the economist explained, "Of course, it was not at [all] surprising that officials knew when we got back and that they had our phone number: we were required to provide this in a health declaration provided prior to departure. My surprise was the efficient way the information provided was used and the friendliness displayed to us by the official in question."

5. Interested readers can learn more about the contents of Resolution 52-NQ/TW, signed on September 27, 2019, to advance the Fourth Industrial Revolution in the *Vietnam Law and Legal Forum*, https://vietnamlawmagazine.vn/party-determines-to-embrace

-fourth-industrial-revolution-16916.html. Decision 749/QD-TTg, on the other hand, signed on June 3, 2020, focused on promoting national digitalization, with the goal to rank among the top fifty countries in e-government by 2030.

6. During the pandemic, policies prioritizing lives over the economy were evident in decisive actions like closing international borders, shutting nonessential businesses, and instituting lockdowns (see act 2). Despite the negative impact on their livelihoods, citizens overwhelmingly supported these early measures. A survey conducted by the Mekong Development Research Institute (MDRI) in cooperation with the United Nations Development Programme (UNDP) in September 2020 reported that 89 percent of respondents agreed it was the right course of action; see https://mdri.org.vn/publications/citizens -opinions-of-and-experiences-with-government-responses-to-covid-19-pandemic-in -vietnam-findings-from-a-phone-based-survey-2/.

7. The declaration, administered by the National Steering Committee for COVID-19 Prevention and Control in coordination with the Ministry of Public Health, was mandatory for entry (and exit) clearance. At the time of my entry, before the borders were closed, I was not required to select a quarantine facility or to upload negative COVID-19 test results, requirements that were later imposed on arrivals with special permission (like invited experts).

8. The "Demystifying Rural Vietnam" study, a collaboration between GroupM, Kantar, and Facebook, surveyed 4,500 rural households to assess social media usage and e-commerce potential in underserved populations. The findings revealed that the internet had overtaken television as the primary medium across all age groups and geographic regions. See www.groupm.com/groupm-vietnam-publishes-research-demystifying-rural -vietnam/.

9. Only about 10 percent of Vietnam's plastic waste is recycled, with the remaining 90 percent released into the environment, including the oceans, making Vietnam one of the top five marine plastic polluters in the world (Liu, Nguyen, and Ishimura 2021, 422).

10. On waste infrastructures and the gendered division of labor involved with recycling and trash collection in Vietnam, see Nguyen (2016) and Schwenkel (2019).

11. Such regulations were also intended to discourage unethical market practices like price gouging or the resale of high-demand products at exorbitant prices on the open market during a health crisis.

12. In Vladimir Sorokin's ([1983] 2008) satirical novel *The Queue*, people spend their days in line to purchase unknown items. This perception of socialist state bureaucracy was exemplified by the Soviet bread line, which, according to Melissa Caldwell, became an "internationally recognized symbol" of the deficiencies of state socialism (2009, 1–2). This resilient image has retained its persuasive power across time; more recently, media coverage of Venezuela has consistently focused on queues as a symbol of economic crisis and socialist state breakdown.

13. Two weeks later, the lockdown would drive these vendors from the streets. While many itinerant traders eventually returned to the city, baguette sellers quickly became obsolete with the opening of new neighborhood bakeries.

14. Readers should note that at the time, there was no mask mandate in the United States, unlike the requirement in Vietnam. In fact, US Surgeon General Dr. Jerome Adams actively discouraged mask use, arguing that they were not effective in preventing viral

transmission among the general population. Official guidelines recommending that people wear masks were not established until April 3.

15. Ratified on February 28, 2020, Resolution 20/NQ-CP allowed the export of masks by license for humanitarian purposes only. It was rescinded two months later on April 29.

16. While global supply chain disruptions during the first year of the pandemic had minimal impact on the everyday consumption of domestically produced goods in Vietnam compared to the United States, certain *imported* items gradually became harder to find, such as specialty flours used for baking breads at a popular French café. Consequently, my idea to purchase imported products to minimize impact on local supplies proved misguided and reflected my own assumptions about scarcity.

17. In Vietnamese: "Để ủng hộ phòng, chống dịch bệnh COVID-19 Quý vị hãy soạn tin: CV n gửi 1407 (Trong đó, n là số lần ủng hộ 20.000 đồng). Mỗi tin nhắn Quý vị đóng góp 20.000 đồng nhân n lần (trong đó n giới hạn từ 1 đến 100). Điện thoại hỗ trợ: 19001530 nhánh 6 (1.000 đ/p). Website: 1400.vn. Xin cảm ơn."

18. In Vietnamese, *truyền thống uống nước nhớ nguồn*. For the portal's mission statement, see https://1400.vn.

19. These statistics, also for other campaigns, can be found at: https://1400.vn/ket-qua-chuong-trinh?tabActive=chuongtrinh. The COVID-19 vaccine fund similarly surpassed all other campaigns in 2021. Notably, the 1407 CV*n* campaign in 2020 averaged sixty thousand VND (US$2.60) per donor. This is particularly significant given that people typically maintained limited cash reserves on their phones at the time due to low telecommunication fees, usually topping up their accounts with twenty-thousand-VND or fifty-thousand-VND cards. (As of 2024, physical cards have become mostly obsolete, and top-ups are now conducted electronically.)

20. The Gold Week campaign (Tuần Lễ Vàng) during the Vietnamese Revolution offers a historical and cultural precedent for this solidarity economy. In 1945, Hồ Chí Minh called on citizens to donate their gold assets in support of independence, establishing a "moral equivalence between people who sacrifice their personal wealth, and those who sacrifice their lives" for the cause (Truitt 2013, 26–27). An estimated 370 kilograms of gold were collected according to the Vietnam National Museum of History. During the pandemic, the redistribution of personal resources was encouraged as mutual aid, but reports of coercion in some instances exposed the underlying power dynamics of such collective efforts. While the media celebrated virtuous citizens donating their pandemic assistance to those in greater need, some village officials allegedly compelled (*bắt buộc*) struggling individuals to redistribute these payments, thereby upholding a narrative of self-sacrifice and solidarity.

21. In Vietnamese: "Thực hiện phương châm 'tạm dừng đến trường, không dừng học,' Bộ GD&ĐT đã công bố tinh giản chương trình học kỳ 2 và hướng dẫn dạy/học qua internet, trên truyền hình. Mỗi thầy cô, gia đình và toàn xã hội hãy chung tay hỗ trợ, động viên, giúp đỡ học sinh học qua internet, truyền hình đảm bảo thuận lợi, chất lượng, hiệu quả."

22. This fitness routine was covered in the national press under the headline "Quarantined Residents in Trúc Bạch 'Fight COVID-19' by Exercising" (X. M. 2020).

23. For insights into the relationship between disease and colonialism in Hanoi, see Vũ Trọng Phụng's novel *Lục Xì* (2011), a satire about sexual health and prostitution. Malarney's (2011) introduction to the book offers valuable historical context. Additionally, Vann

and Clarke's (2018) graphic novel offers a creative depiction of the battle against rats and the plague in the colonial city.

24. While available electronically, these copies had been printed out for me to read and sign. In Vietnamese, this paperwork was titled as follows: (1) "Quyết định 110/QĐ-UBND về việc áp dụng biện pháp cách ly y tế phòng, chống dịch bệnh COVID-19"; (2) "Hướng dẫn cách ly tại nhà/nơi lưu trú; (3) Bản cam kết thực hiện các biện pháp cách ly y tế tại nhà để phòng, chống bệnh viêm đường hô hấp cấp (COVID-19); and (4) "Những điều cần biết để phòng bệnh viêm đường hô hấp cấp do vi rút corona mới."

25. Neighborhood leaders, mostly retirees, serve as the eyes and ears on the streets. During the pandemic, they were responsible for enforcing lockdowns and organizing food and medicine for families in need. Largely unrecognized essential workers, they are considered the "last line" (*tuyến cuối*) of the government apparatus, the "forces" (*lực lượng*) on the ground closest to the people (Thái 2021).

26. The texts followed the same general structure and typically read:

Bộ Y tế đề nghị Quý vị thực hiện tốt 5 điểm sau đây: [The Ministry of Health recommends that you implement the following 5 measures well:]

1. Hạn chế tối đa ra ngoài, chỉ ra ngoài khi thực sự cần thiết. [Limit going out as much as possible, only go out when absolutely necessary.]
2. Nếu buộc phải ra ngoài luôn đeo khẩu trang, hãy giữ khoảng cách tiếp xúc, tốt nhất là 2m. [If you must go out, always wear a mask, maintain distance, preferably two meters.]
3. Thường xuyên rửa tay bằng xà phòng hoặc dung dịch sát khuẩn. [Wash your hands often with soap or an antiseptic solution.]
4. Vệ sinh và để thông thoáng nhà cửa; lau rửa thường xuyên các bề mặt, các điểm hay tiếp xúc; sinh hoạt lành mạnh. [Clean and ventilate the house; frequently wipe down surfaces and high-contact areas; maintain a healthy lifestyle.]
5. Thực hiện khai báo y tế trên ứng dụng NCOVI (ncovi.vn), hoặc khai trực tuyến trên tokhaiyte.vn; cập nhật tình hình sức khỏe hàng ngày, giữ liên hệ với cơ sở y tế. [Make medical declarations on the NCOVI app (ncovi.vn) or report online at tokhaiyte.vn; update health situation daily, keep in touch with health care facilities.]

27. Despite Vietnam's aggressive screening strategy, state laboratories struggled to process the volume of samples. When I inquired about my airport test results, Hương attributed delays to overwhelmed facilities. This surge followed the second major outbreak in Hanoi on March 18 at Bạch Mai Hospital, which affected food service personnel, and occurred nearly two weeks after the Trúc Bạch Street lockdown. Lower-risk cases like mine were not prioritized for processing, she informed me.

28. Vietnam's distinctive blue protective gear became a marker of Vietnamese identity in airports. All Vietnamese nationals on repatriation flights were required to don these disposable, full-body suits before boarding and keep them on until arrival at collective quarantine facilities on army bases, where they stayed in barracks for two weeks at no cost. I wore the blue suit myself in October 2021 traveling from Hanoi's Nội Bài Airport to hotel quarantine (where foreigners were housed at their own expense) and then again from the hotel to my apartment, accompanied by nurses. While presumably made of breathable material (non-woven polypropylene), the suit felt stifling if worn too long or in unventilated

conditions, a common complaint among returnees that I also experienced. To illustrate the strict entrance regulations for returning nationals at the time, a woman who discarded her suit after deplaning in Đà Nẵng due to thermal discomfort was told she would not be allowed into the country. Fortunately, my student who was with her and later recounted this incident to me gave her a second suit she had stuffed into her carry-on as an ethnographic keepsake.

29. The Bạch Mai outbreak reinforced the perception that hospitals were places of danger and death to be avoided rather than sites of healing and recovery. This view led to concerns among authorities that people might conceal their symptoms, as hospitalization was mandatory for anyone displaying signs of illness at that time.

30. Dispatch 16212/BYT-MT from August 2, 2021, advised authorities against the indiscriminate spraying of chemicals in outdoor areas or on individuals, including through "disinfection booths" (*buồng khử khuẩn*), unless specifically addressing confirmed COVID-19 cases. Despite this guidance, the practice persisted. For example, during the opening weekend of Hanoi's metro (sky train) in November 2021, security sprayed disinfectant on my clothing and hands before I was permitted to enter the station.

31. The pledge read, "Gia đình chúng tôi xin cam kết thực hiện đúng các nội dung trên, nếu không tôi và gia đình xin chịu mọi trách nhiệm theo quy định của pháp luật" (Our family commits to comply strictly with the above measures, otherwise my family and I will bear all responsibilities according to the provisions of the law).

32. The term "sonosphere" has been used to describe human and nonhuman engagement with the "sonic envelope of the earth" (Oliveros 2011, 162).

33. Nancy distinguishes between listening (the attempt to make sense or meaning beyond sound) and hearing (the outcome or understanding of that perception): "To be listening is always to be on the edge of meaning" (2007, 6).

34. Our digital relationship did not end with the emoji text. A week later, Hương sent an urgent message: "Chị ơi [Hey older sister/senior female person], I just received your test result. It's negative chị, ok!" After I expressed my thanks for the follow-up, she replied, "Hee hee. Wishing you happiness always, chị." I responded in turn with a smiley emoticon, concluding our exchange.

ACT 2: LOCKDOWN

1. Kate Lacey (2013) examines how mediated listening practices shape public life by fostering "listening publics" that fluctuate between engaged attention and deliberate disengagement, highlighting the dynamic nature of public participation in an era of diversified media.

2. For a ratified version of Directive No. 16/CT-TTg, see https://vncdc.gov.vn /mediacenter/media/files/1012/08-2021/805_1629106225_571611a30316f431.pdf. Additional COVID-19 directives are also available on the Ministry of Health (MOH) website: https:// moh.gov.vn (in English: https://moh.gov.vn/web/ministry-of-health). To ensure information accessibility during the pandemic, the MOH homepage provided twice-daily updates on positive test results and rates of infection by locality, as well as ongoing nationwide tallies of individuals tested and quarantined.

3. Health authorities identified Patient 91, a Scottish pilot working for Vietnam Airlines, as the source of the outbreak at the popular expat establishment. Suspicion of foreigners

as potential carriers of the virus escalated when the Buddha Bar became the country's first instance of "serial transmission" (*chuỗi lây truyền bệnh*) and site of the largest COVID-19 cluster in southern Vietnam at the time. The bar also drew condemnation from the local sangha for appropriating Buddhist imagery for non-sacred, commercial activities (causing it to later rebrand itself as the Nameless Bar).

Patient 91 would remain at the center of national attention for the next several months as his lungs collapsed and Vietnamese doctors, with the support of a global advisory committee, extended tremendous effort and resources to keep him from becoming the country's first pandemic mortality. The pilot's unexpected recovery—another COVID-19 "success" story—was a source of great pride among many Vietnamese on social media, some of whom volunteered to donate a lung to save him. At the same time, his story "unsettled care" by exposing how the politics of caretaking continue to be shaped by western patriarchal privilege (Murphy 2015, 721). The pilot's reveling, jet-steering, white male body did not attract the same scorn as the reveling, jet-setting, Vietnamese female superspreader (see pandemic figure 1). Nonetheless, the high cost of his long-term care did draw critical commentary about the disproportionate distribution of health care resources.

4. This moderate approach would change a year later with the rapid spread of the highly contagious Delta variant, which necessitated stricter lockdowns (see coda).

5. Although low compared with other areas of the world, Vietnam's documented COVID-19 cases reached 212 on March 31 when the directive was issued, representing a more than twofold increase over ten days from the 94 infections reported on March 21. This rapid rise prompted authorities to take bold measures to stop the threat of further spread.

6. To provide readers with a sense of the global scope of the pandemic at the time, on March 20, 2020, while Vietnam had recorded only ninety-one coronavirus cases cumulatively with no deaths (an increase of six infections over the previous day), Italy experienced the highest death toll in the world, with 627 fatalities in a single day.

7. "*Ở nhà là yêu nước*" was one of the popular slogans that the state used to encourage people to comply with the lockdown. In response to Directive 15, the Hanoi Youth Union launched a photo challenge on its Facebook page on March 28, 2020, only days before the lockdown began. The challenge garnered widespread participation, including TikTok videos, from all sectors of society. From infants to the elderly, health care workers, students, police officers, politicians, and celebrities, people posed for the camera holding signs with short rhymes, pledging to stay home to care for themselves and their families in a show of national unity.

8. The utterance "*chống dịch như chống giặc*," which quickly evolved into a soundbite, slogan, and hashtag, was first made during the prime minister's broadcasted speech on January 27, 2020 (the third day of the lunar new year). The speech announced the formation of a National Steering Committee and the launch of a whole-of-government response following the first reported COVID-19 cases linked to travelers from Wuhan, China, on January 23. Schools would not reopen after the Tết lunar new year holiday. https://vpcp .chinhphu.vn/thu-tuong-chu-tri-cuoc-hop-ve-phong-chong-dich-ncov-11523170.htm.

9. For the April 2020 V-pop song and video "Chống Dịch Như Chống Giặc," featuring musician Ưng Đại Vệ alongside a dancing mouse doctor, see www.youtube.com /watch?v=AxfMawng1Fk. For the children's exercise routine inspired by the song to promote public health, see https://www.youtube.com/watch?v=Wb25epUnhZw&list=RDWb 25epUnhZw&start_radio=1.

10. "Mỗi người dân hãy là một chiến sỹ phòng, chống dịch; tiếp tục chung sức, đồng lòng đẩy lùi dịch bệnh." A ratified copy of Directive 15 is available on the Ministry of Health's website, https://moh.gov.vn/documents/176127/356256/27.3.2020+CT+15+CT -TTg.pdf/9c07d0c0-3bde-4003-a605-786b752f335c.

11. Unlike cities in the United States and Europe, where collective applause for first responders and frontline workers became a sonic feature of urban life, there was no communal clapping during lockdown in my neighborhood.

12. In the early months of the pandemic, when infection rates were low and the spirit of solidarity high, this out-migration did not spark any widespread concern or controversy. Later in 2020, as villagers became weary of urbanites fleeing outbreaks to "safer" grounds, provincial authorities required those arriving from the city to quarantine for seven days. This mandate also applied to migrant workers returning home from pandemic-affected areas.

13. A large number of expatriates ("expats") also left for their home countries before international borders closed, seeking to be closer to their families. Over the next months, I would see very few foreigners in Hanoi.

14. Self-barricading *in* to keep the threat of the Other/Unknown/Uninvited *out* was a common spatial strategy of care and protection used in Hanoi. Alleyway gates were closed, doors to stairwells in collective housing (*khu tập thể*) locked, and entries to markets blockaded, altering forms of sociability as people were forced to interact and transact across protective barriers that fortified an architecture of separation across the city.

15. Wartime evacuees, for example, were sent to remote locations far from the city, often in mountainous areas, to escape US bombing raids on urban infrastructure. See chapter 2 in Schwenkel (2020).

16. This "imaginary of rural immunity," and its ethnic and class distinctions, is not unique to Vietnam. On the ways that rural exceptionalism masked the racialized dynamics of coronavirus risk in the United States, see Sangaramoorthy and Benton (2020).

17. For example, when I traveled to Lào Cai Province in July 2020 during the period of spatial distancing, local Hmong residents told me that there had been no need to stay at home during the nationwide lockdown, since the virus was only present in lowland cities, implying a disease of urban (and Kinh) "civilization."

18. For more on the "*bỏ phố về rừng*" (leave the city for the forest) movement, which is also a Facebook group, see Long Nguyen (2020).

19. On the "nocturnal turn" in anthropology, see Galinier et al. (2010) and the special issue of *Ateliers d'Anthropologie* (Becquelin and Galinier 2020).

20. As Feld (1996, 97) has argued, sound is integral to how we sense, come to know, and experience place as "sensual space-time."

21. The original Vietnamese version of the section epigraph is: "Loa phường là sức mạnh của chính quyền, là sợi dây nối giữa chính quyền và dân." In this 2017 declaration, General Bạch Thành Định reaffirmed the power of loudspeakers to embody state power and foster social cohesion, arguing for their continued relevance in the age of the Internet.

22. Municipal regulations restrict broadcasts in central urban districts to two fifteen-minute periods per day, Monday to Friday only, unless a state of exception, such as a public health emergency, has been declared. However, this regulation is generally not enforced and has been widely ignored.

23. According to a municipal official, during the pandemic, the Department of Information and Communications issued a directive encouraging increased frequency of loudspeaker broadcasts. While my ward maintained the standard twice daily transmissions with occasional supplemental announcements using other auditory technologies, adjacent wards chose to broadcast as often as every few hours.

24. Loudspeaker deactivation depended on urban location. While the loudspeakers in my ward (Trúc Bạch) were typically silent on any given day prior to the pandemic, across the lake in Quán Thánh Ward authorities broadcasted more regularly, typically beginning at 6:15 a.m. and often ending with classical music.

25. One counterargument that emerged in the "keep or abolish" (*bỏ hay giữ*) debate is that real-time digital platforms like Zalo are able to transmit urgent information more quickly than loudspeakers, which, although faster than print media, still requires preparation time. For differing positions on the debate, see Lê Thanh Tâm and Lưu Ngọc (2022).

26. For example, electroacoustic technologies, like headphones, enable users to create, curate, and isolate themselves in private auditory worlds (Sterne 2003, 87). However, this does not suggest a gradual individualization of listening publics. As Lacey (2013) argues, listening has historically co-existed as both collective and private experience.

27. The revolutionary song referenced in the epigraph highlights how this ethos of sonic socialism is profoundly gendered. Loudspeaker broadcasting in Vietnam has often been performed by women as a form of affective labor intended to alleviate uncertainty by projecting a sense of security. This emotionally resonant, maternal voice of care recalls figures like Hanoi Hannah, whose wartime broadcasts combined intimacy and threat to unsettle American GIs. Beyond psychological warfare, the female voice emerged as a revolutionary symbol—nurturing yet commanding—in poetry and song.

"Listen to the Voice of the Homeland" ("Lắng Tiếng Quê Hương") was originally written for Voice of Vietnam radio by composer Dân Huyền during the US war in Vietnam. It was later augmented with this verse, by singer Kim Cúc (Ó Briain 2022, 91). In Vietnamese the verse reads:

> Nỗi vui mọi nhà ấm trong những tiếng loa
> Tiếng nói ngọt ngào và tiếng hát thiết tha

28. Vietnamese elites characterized illiteracy under colonialism as a disease "as serious a threat as cholera," according to David Marr (1981, 178). Upon becoming president in 1945, Hồ Chí Minh launched a nationwide campaign to promote universal literacy in newly independent Vietnam under the rallying cry "Studying Is Patriotic" (1981, 184–86), a slogan echoed seventy-five years later in the COVID-19 campaign, "Staying at Home Is Patriotic." Remarkably, by 1958, illiteracy among the lowland population aged twelve to fifty had dropped to just 6.6 percent, a dramatic decline from 93.4 percent in 1945 (1981, 187).

29. Like French Indochina, French Algeria witnessed a radical transformation of radio from a tool of colonial domination and symbol of bourgeois consumption into a weapon to achieve liberation through broadcast communication (Fanon 1965, 89).

30. Despite a few iconic exceptions that gained international attention, large COVID-19 posters displayed in public spaces in Hanoi were relatively uncommon during this period. Far more prevalent, yet less conspicuous, were smaller reproductions of fourteen selected public health graphics attached to the external walls of People's Committees, ward- and

commune-level health stations, or other public buildings. These resulted from a March 2020 Arts Council competition that drew 103 submissions from twenty-three artists. The Eurowindow emblem featured on these posters illustrates how certain commercial entities aligned themselves with public health efforts to bolster the solidarity economy, akin to the rice and mask ATMs discussed in act 1. This joint-stock company, founded by a Vietnamese billionaire trained in socialist Eastern Europe, both funded the competition and printed the selected designs, exemplifying how private enterprise collaborated with state initiatives through strategic volunteerism.

Readers will note that I deliberately avoid using the term "propaganda" poster, given its negative connotations of ideological manipulation and misinformation. The Vietnamese term *tuyên truyền* more neutrally signifies what the Global North calls public awareness campaigns, often negatively labeled "propaganda" when originating from the Global South. For a deeper analysis of this perspective in the context of war photography, see Schwenkel (2009b).

31. There was no fixed narrative for Vietnam's pandemic timeline. Initially, the state described the first year of COVID-19 as unfolding in three distinct phases, each defined by a pivotal *outbreak*. Phase 1 began on January 23 with the first reported cases of imported infection; over the next six weeks, a total of sixteen cases were recorded, all of whom fully recovered. Phase 2 started on March 6 with Patient 17, following twenty-two days without reported infections nationwide. The emergence of community transmission shifted perception of the virus from a foreign threat to a domestic challenge, necessitating a multi-faceted prevention strategy. Phase 3 commenced on July 25 with the outbreak in Đà Nẵng, after ninety-nine days without community transmission. The first death was recorded on July 31. A later timeline proposed by the Ministry of Health was organized according to *variants*: Phase 1 (starting January 23) featured the initial strain of SARS-CoV-2, presumed to have originated from Wuhan, China; Phase 2 (starting July 25) involved the D614G variant; Phase 3 (starting January 28, 2021) was marked by the Alpha variant; and Phase 4 (starting April 27, 2021), the deadliest, was dominated by the Delta variant.

32. Sound can act as both a response to and a solution for crisis. Engineering professor Markus J. Buehler's research on sonification has attempted to convert the vibrational structures of coronavirus protein chains into music. These structures, he argues, are "too small for the eye to see but they can be *heard*" (emphasis added), potentially facilitating drug design (International Science Council 2020).

33. Readers may experience a sense of déjà vu when encountering this loudspeaker script, as it echoes the hygiene code of conduct I signed upon my quarantine in act 1, as well as the recommended pandemic protocols mass-texted by the Ministry of Health. The state's public health messaging, particularly its audio component, relied heavily on repetition across various media to establish biomedical truths as common sense, causing some people to gradually disengage due to message fatigue.

34. The *bảng thông báo* can be considered a complementary technology to loudspeakers, both serving as essential tools of urban governance to facilitate state-citizen interaction (Schwenkel 2020, 225–27). As a reminder of the layered structure of urban governance: wards in Hanoi are third-tier administrative units under districts (*quận*) and centrally controlled municipalities (*thành phố trực thuộc trung ương*) but above residential clusters or *khu phố*. They are the lowest level of government responsible for implementing state policy and managing citizen-state relations (Koh 2006, 32).

35. In the early months of the pandemic, when case numbers were small and manage-able, the state deployed a vast team of thousands of contact tracers tasked with mapping the detailed mobility history of each confirmed infection (F0) and their direct and indi-rect contacts, extending to F5s. This exhaustive history, including exact addresses and the specific days and times when F0s visited particular locations (which then had to close), was communicated to the public as part of a widespread effort to trace and contain all potential persons who may have inadvertently encountered the virus.

36. Written by Xuân Hồng, the song captures the rhythmic sound of Xtieng women in the Central Highlands pounding rice at night (“*cum cụp cum, cum cụp cum*”), preparing food for soldiers en route to the front.

37. The collective consumption of audio and audiovisual media, including urban radio and rural mobile cinema, extends to print media as well. For instance, newspaper publish-ers would display daily editions in glass boxes outside their headquarters for communal reading. In Hanoi, it was not uncommon to see groups of readers, particularly older men, gather to discuss the content. However, the pandemic and digitalization have diminished this social practice in recent years.

38. This broadcast can be accessed at https://vovgiaothong.vn/newsaudio/nguoi-ha-noi -van-muon-co-loa-phuong-nhung-theo-cach-khac-d28098.html.

39. This survey of public perceptions of government responses to COVID-19 can be accessed at https://mdri.org.vn/publications/citizens-opinions-of-and-experiences -with-government-responses-to-covid-19-pandemic-in-vietnam-findings-from-a -phone-based-survey-2/. It followed a May 2020 YouGov poll, which similarly found that 94 per cent of the population trusted the state’s pandemic management, compared to just 43 percent in the United States: https://yougov.co.uk/international/articles/29757 -international-covid-19-tracker-update-18-may. Despite this strong support, the MDRI and UNDP survey found that the majority of households had been negatively impacted by the pandemic, especially those with workers in the informal sector, due to job and income losses.

40. The MDRI and UNDP survey revealed a striking 99 percent mask compliance rate among respondents.

41. For example, a *Deutsche Welle* article described Vietnam as “winning its war” against the coronavirus not through medical and technological solutions, or even neigh-borhood solidarity, but through a dystopian surveillance state with “Communist Party spies” allegedly monitoring every corner, creating an atmosphere of mutual suspicion that placed immense social pressure on people to comply (Ebbighausen 2020).

42. The lockdown initially transformed the permeable boundaries between public and private life into more rigid divisions as people retreated inside their homes. By the second week, these boundaries began to blur again, albeit cautiously.

43. In her analysis of nuisance lawsuits targeting airport noise in Los Angeles, Marina Peterson observes how noise brings into sharp relief the historical relationship between sound, property values, and ownership rights, such as the right to quiet enjoyment. She notes that the legal foundation of noise in nuisance law has been “concerned principally with the protection of private property” (2021, 158).

By contrast, in my neighborhood in Hanoi, noise associated with private construction complicated assumptions about whose rights were being protected—or exercised. Here, the rhythms of construction aligned more closely with informal labor practices than with noise ordinances. Migrant workers, who constitute the majority of construction teams, live on-site, seldom leave the premises, and work from dawn to dusk, seven days a week. Noise complaints are instead settled privately. For instance, when early morning demolition and jackhammering next door made my apartment unlivable in 2022, the issue was resolved with a monthly compensation of two million VND paid by my landlord (though it may have come from the homeowner—this was unclear). Breaches of building regulations are often accommodated with a "fine to exist" (*phạt cho tồn tại*) or other unofficial payments, as authorities turn a blind eye to noise and related violations, sometimes out of sympathy for families seeking to improve their living conditions (Schwenkel 2020, 275).

44. The 2010 National Technical Regulation on Noise (QCVN 26: 2010/BTNMT) sets the maximum noise limit in residential areas at 70 dBA from 6 a.m. to 9 p.m. and 55 dBA during nighttime hours.

45. Asking for sympathy, *xin thông cảm*—literally a plea (*xin*) for the transfer (*thông*) of feeling (*cảm*)—is a culturally resonant expression that seeks an emotional connection and concession when facing adversity or circumstances requiring flexibility in formal government or informal social settings.

46. For example, Nguyễn (2022) and Kim (2015) highlight informal negotiations between street vendors, homeowners, and local police in response to sidewalk clearance campaigns that disproportionately impacted rural migrant women, who were often met with sympathy from residents and ward officials.

47. On the sounds of Black sonic dissent inspired by the Black Lives Matters movement, see also Brooks (2016).

48. For instance, during spatial distancing (post-lockdown), I would descend to the ground floor, open the back door, and call out what I needed, without leaving the premises. A mere few feet away, Giang would point to different items spilling out from her narrow entryway onto the alley, encouraging me to purchase more. When she handed me the bag with my order, I handed her cash. Many of my building's residents secured food this way without having to venture outside. As the pandemic eased in 2022, Giang turned from selling produce—which was no longer profitable after street vendors returned, she told me—to selling *bún đậu* (rice noodles with fried tofu), demonstrating her household's flexible livelihood strategy to manage precarity.

49. For example, while I was writing this chapter on a clear and sunny winter day in Hanoi in January 2023, the air quality index (AQI) in my lakeside neighborhood was coded red at 154 ("unhealthful"), prompting warnings to avoid outdoor exercise. The fine particle matter (PM 2.5) index reached an alarming 61.5, more than twelve times the WHO's recommended limit of 5.

50. Periodic roadblocks and checkpoints along the main Hanoi-Hải Dương thoroughfare prevented migrants from either reaching the city or returning home, as was the case with the security guards from my building. Serving as a social control mechanism, state media, particularly newspapers and television, highlighted instances of individuals fined for violating the ban on travel across administrative boundaries.

51. The proximity of my neighborhood to government offices and ministries, including the National Assembly, has resulted in a high concentration of officials and civil servants living there. This demographic distribution is rooted in a history of land and housing allocation, which clustered cadres in Trúc Bạch, among other wards. However, the examples I have provided demonstrate sonic dissent across social strata, ranging from an economically

vulnerable family engaged in vending (historically viewed with suspicion) to a more well-off family with possible connections to local authorities.

52. See, for example, Ha An (2020, 24). The day before the stay-at-home order went into effect, delivery orders surged by 91 percent, with average order value increasing 26 percent as consumers stocked up on food supplies for their families. In the first half of 2020, food delivery experienced a year-over-year increase of 1,800 percent, leading to tens of thousands of daily deliveries and a tremendous rise in plastic waste, much of which ended up in the ocean. See also act 1, note 9.

53. Vietnamese health authorities distinguished between imported cases (*ca nhập cảnh*) and cases of community transmission (*ca trong cộng đồng*), which were of greater concern due to the unknown source of origin and potential for rapid and widespread contamination before detection through contact tracing. During these three months, approximately 150 imported cases were reported, primarily from repatriating Vietnamese citizens and experts arriving on chartered international flights, with infections detected while in centralized quarantine. There were also cases linked to illegal border crossings from China and Cambodia, likely exceeding official figures.

PANDEMIC FIGURE 4: THE GRAB DRIVER

1. They also won out over new local players like GoViet (turned Gojek in 2020) and Be, which became competitive with GrabBike and informal *xe ôm* motorbike taxis, but not GrabTaxi in Hanoi.

2. This discourse of freelance capitalism and drivers as self-employed "partners" (*đối tác*) rather than "workers" can be viewed at https://www.grab.com/vn/.

3. Resolution 42/NQ-CP, ratified on April 9, 2020, by the prime minister, provided cash assistance to persons experiencing a loss of income or other COVID-19 difficulties. The 62.2 billion VND social protection package applied to those with and without formal labor contracts, meaning informal (usually migrant) workers were also recognized, but only in the place of their household registration. The relief package seemed to exacerbate rather than alleviate vulnerability, which was a source of debate. Registration procedures were complex and reports of pressure from officials to regift the cash payment emerged in the provinces, pointing to a shadow COVID-19 economy. This transpired despite the resolution clearly banning "policy profiteering" (*trục lợi chính sách*). In Hanoi, cash transfers occurred at the ward level, with criteria that appeared to vary by jurisdiction. For example, in Cửa Nam Ward of Hai Bà Trưng District, all registered residents claiming hardship, including nonnationals, received one million VND (US$43) in cash, with the possibility of regifting to less fortunate neighbors. However, such widespread distribution was not the case in my locality.

4. A pseudonym to maintain confidentiality.

5. Despite reduced taxes and fees, car prices in Vietnam remain significantly higher than those in other Southeast Asian countries, often twice as high. When I joked with one Grab driver that his Hyundai was several thousand dollars cheaper in the US market, he quickly countered by calculating the annual registration and insurance costs in the United States, which are substantially higher than in Vietnam. He argued that while the upfront purchase price was steeper, the modest operating expenses meant his vehicle was more economical over the long term.

6. Stories of unauthorized border crossings by infected individuals were commonly reported along both the Chinese and the Cambodian borders, with subsequent disease outbreaks attributed to these breaches.

ACT 3: SOCIAL DISTANCING

1. On social distancing as an altruistic stance, if not an "empathetic response based on the recognition of mutual needs," see Brakman (2020).

2. Note that the term "social distancing" was not widely used in Vietnam in 2020. Instead, the general term "distance" (*giãn cách* or *khoảng cách*) was employed to emphasize the need to maintain *physical* space between bodies. The preference for "social" rather than "spatial" distancing in the United States and elsewhere has been criticized by health care professionals for conveying a sense of disengagement from social relationships.

3. In Vietnamese:

> Ta cùng nàng nhìn nhau không tiếng nói
> Sợ há mồm vi khuẩn bắn sang nhau
> Đôi hơi thở thì thôi ta cố nín
> Đôi linh hồn chìm đắm bể u sầu

Poetic adaptations attributed to Thầy Thanh Sơn. "Thơ tình thời Covid," Facebook, August 5, 2020, www.facebook.com/1034587036674551/pcsts/2071411646325413/.

4. As a reminder, individuals identified as Fos—those with confirmed infections—were hospitalized, while F1s (those who had contact with Fos) were placed in centralized quarantine for a minimum of two weeks, pending test results. In contrast, F2s and F3s were allowed to quarantine at home. However, if traveling, an on-site quarantine was required, heightening concerns about being too far away in the event of an outbreak.

5. This shift in public sentiment became evident through readers' online comments on cases covered by the media and was influenced by factors such as race and gender. Generally, people extended their good wishes to those afflicted by the coronavirus, especially foreigners, with whom they were more forgiving (see the case of the Scottish pilot identified as Patient 91, discussed in act 2, note 3). However, sympathy could quickly turn to outrage and stern rebuke when breaches of protocol occurred, showing how public care was tied to the display of a proper ethical disposition. An illustrative case involved an asymptomatic male Vietnam Airlines flight attendant—Patient 1342—who in November 2020 disregarded mandatory home isolation following his return flight from Japan. His casual flânerie around the city, with each location and time documented by the press to locate F1s–F4s, set off a new chain of infections, ending an eighty-nine-day streak without community transmission and resulting in the quarantine of two thousand individuals. In March 2021, the flight attendant was handed a two-year suspended jail sentence in the first instance of legal action against a person for violating COVID-19 regulations. On his discriminatory treatment by Netizens owing to speculations about his sexuality following details of his urban pastimes, including people visited and places frequented, see Vũ (2020).

6. Cities across the globe, particularly Asian megacities, experienced a bicycle boom during the pandemic with the suspension of public transport systems. In my ward, the surge in interest was more for recreational than commuting purposes. Like the feminist bicycle collective in Quito discussed by Julie Gamble (2017), women cyclists in Hanoi made

claims on urban space and to the use of city streets through group cycling. Unlike in Quito, however, cyclists in Hanoi, myself included, faced a lack of urban infrastructure, amplifying the risks of riding while also rendering bikeable lakefront neighborhoods like mine more inviting.

7. Here I differentiate public fitness (group exercises) from team sports according to their level of competitiveness. The former tends to strive for collective wellness through synchronized physical activity, while the latter places greater emphasis on friendly rivalry in pursuit of a particular outcome or victory. Although gendered, these distinctions are not absolute; women, for example, commonly play on badminton teams, and while it is rare to find men in aerobics classes, they do participate in ballroom dancing. Other competitive pursuits commonly observed in public parks, such as late afternoon matches of *cờ tướng* (*xiangqi* or "Chinese chess"), are typically associated with male activities and displays of camaraderie. Both men and women walk, bike, and use the outdoor fitness equipment installed by ward authorities.

8. Although walking is a practice found across cultures, the history of walking is not equally "everyone's history," as Solnit suggests (2001, 4). Instead, it is a history marked by privilege and the policing of certain walking bodies, like Black bodies and women's bodies, as scholars have shown. On the risks of "walking while Black," see Cadogan (2016). His experiences of walking in the United States, which he describes as a "complex and often oppressive negotiation," offer a somber counterpoint to celebrations of walking as liberating. The aimless strolling and urban wandering attributed romantically to flâneurs—white affluent men who enjoyed the freedom of mobility without fear or harassment—were also historically inaccessible to women (Elkin 2016; Wilson 1992).

9. The fines for violating pandemic regulations, posted by the ward's COVID-19 Steering Committee, read:

- Not wearing a mask in public: 1 million VND [or US$43]
- Holding large gatherings: 20 million VND
- Conducting business and services in public places: 20 to 40 million VND
- Improper disposal of used mask in public places: 3 million VND, or on the sidewalk or street: 7 million VND

10. Despite claims that the video might surpass "Ghen Cô Vy" in popularity, "Việt Nam Ơi! Đánh Bay Covid" never garnered significant global attention, amassing only 5.7 million views on YouTube compared with 87.5 million views of "Ghen Cô Vy." Nonetheless, "Việt Nam Ơi! Đánh Bay Covid" had a broader impact on everyday life in ways that I discuss in this chapter. The official video can be viewed at www.youtube.com /watch?v=6AJitnTAB7M. An English-language version released a week later that emphasized global togetherness in the fight against the virus amassed a mere 250 thousand views: www.youtube.com/watch?v=moEpafU5XoQ. Like with "Ghen Cô Vy," the video "Việt Nam Ơi! Đánh Bay Covid" included sign language to incorporate deaf Vietnamese citizens into the state's sonic campaigns.

11. It is not an overstatement to claim that "Việt Nam Ơi!" has become an unofficial national anthem. While waiting at the Vietnamese embassy in Berlin in July 2023, I found myself listening to an audio track of the song playing on the television monitor, accompanied by images of the country's mountainous scenery.

12. On the historical use of amulets during a cholera epidemic in colonial Annam, see Cadière (1910). In conversations about precautionary steps taken at home to prevent COVID-19, urban respondents mentioned tonics (*thuốc bổ*) containing fresh ginger and garlic, also infused with alcohol, as well as recipes like rice porridge with fresh onion and herbs to boost immunity.

13. On sonic agency as an emancipatory practice that reconceptualizes dominant logics of publics and public spheres by "communities in movement," see LaBelle (2018, 94).

14. Ingold has been a vocal critic of the tendency in scholarship to "slice up" the environment along differing sensory pathways or "scapes" (2011, 136), especially understandings of the eyes and ears as "separate keyboards for the registration of sensation" (2000, 268) rather than as part of a cooperating system. For a rebuttal that prioritizes the social and the material dimensions of sensory interactions over the phenomenological, see Howes (2022).

15. For a discussion of urban soundscapes as intangible heritage in Vietnamese cities, see Officine Gặp's research project *Crafting a Sonic City* (Rebuscini and Frassi 2022). In this study, respondents lamented the auditory shift in hawkers' tunes from live melodies to recorded tapes, attributing such change to the influence of modernity and its diminishment of sensory experiences.

16. See, for example, the 2017 report by the non-government organization TRAFFIC. https://www.traffic.org/site/assets/files/1580/caged-in-the-city.pdf.

17. This predominantly male sport reemerged quickly and with minimal risk, due to players being spaced several meters apart. Similar to parkour, *đá cầu* incorporates the built environment into its practice. For instance, around Trúc Bạch Lake, a group of middle-aged players repurposed a police barrier as a makeshift net. Such creative adaptation is not uncommon. People on foot frequently integrate elements of urban infrastructure into their exercise routines, using benches for reverse sit-ups or guardrails for stretching.

18. Though they lack the vocal capacities of birds, fish also communicate with one another through both active and passive sounding. On fish sound production research and its inventory of underwater recordings, see https://fishsounds.net/. On transductive ethnography and the anthrophony of deep-sea soundscapes, see Helmreich (2007).

19. For a personal account of one family's survival from unlawful fishing in West Lake during the war and postwar subsidy period, see L. Phạm (2020).

20. West Lake's surface area, approximately five hundred hectares, is fifty times that of the surface area of Trúc Bạch Lake, making its freshwater habitat considerably more diverse and attractive to organized poachers. It takes around one hour to bike West Lake's perimeter of eleven miles with all its twists and turns (and traffic), compared with ten minutes to cycle around Trúc Bạch Lake (1.6 miles).

21. Also known as Cẩu Nhi Temple. Built in the eleventh century, the origins and purpose of this temple—to worship dog spirits, the water goddess (Mẫu Thoải), and possibly fish—remain a subject of ongoing historical debate and point to shifts and overlaps in the objects of animist worship across time, as suggested by historian Phan Huy Lê (2018).

22. These kinds of solidarity alert systems are not unique to Hanoi. During a visit to Phú Quốc, for example, vendors let out a series of high-pitched whistles to signal the arrival of police. In Trúc Bạch, while people yelled to alert one another, the sound of approaching bullhorns echoing across the lake also afforded them time to clear the area before police appeared on the scene.

23. Sidewalk checks were usually conducted twice a day, once in the morning, around 7:15 a.m., and then again in the late afternoon, with the possibility of a third check on weekends when cafés were more crowded. Police action unfolded quickly. One afternoon during spatial distancing, as I set out for a 5 p.m. bike ride, a police warning horn blared behind me, startling me with its proximity and volume. "Whose motorbikes are these on the sidewalk?" an authoritative male voice boomed through the megaphone. I stopped to observe as café staff scrambled to clear chairs and tables from the sidewalks while owners rushed to move their vehicles and vendors hurriedly retreated behind the canal. By the time I completed my first loop around the lake just ten minutes later, the vendors had returned, and customers were once again seated along its perimeter, as if no disruption had taken place. On expressions of solidarity between vendors and homeowners, some of whom offer support by hiding traders' merchandise during police checks, see Nguyễn (2022) on Hanoi and Kim (2015) on Hồ Chí Minh City.

24. Unfortunately, these trees were felled during the 2023 Trúc Bạch Lake renovation project, after which the tea vendor and her neighbors did not return. This outcome exemplifies how urban beautification disproportionately harms small-scale, female-led businesses that depend on public space. The redesign also replaced gathering areas used for aerobics classes with bamboo gardens, disrupting women's social and economic activities.

25. The middle-aged women on the main strip transported their makeshift operations—consisting of thermoses filled with hot water, tea leaves, a bucket of ice, drinking glasses, and plastic stacking stools—by motorbike with the assistance of male household members. In contrast, the elderly tea seller transported her equipment, including candy and cigarettes, alone on her bicycle. While the other women closed shop around 7 a.m., some likely continuing to other jobs, this older vendor operated until 3 pm. Afterward, a different female vendor, who catered to groups of men playing cards or games of Chinese chess (cờ tướng) occupied the space, illustrating how gendered soundscapes evolved across the day.

26. Some vendors purchased these recordings in the market, while others digitally created them, using USB to transfer the files to their speakers—another example of sonic media convergence of "old" and "new" technologies.

27. As the musician and sound artist Nguyễn Thanh Thủy observes, the calls of street vendors often follow the rhythmic—and rhyming—structure of traditional oral poetry, including *lục bát* (six-eight) poetic forms (Östersjö and Nguyễn 2016).

28. Nguyễn Trung Kiên's (2022) essay on the history of sports in Vietnam references several iconic images of Hồ Chí Minh participating in physical activity to promote health and revolutionary values, the most famous being a photograph of him lifting weights.

29. Alternatively, "weak citizens make for a weak country" (*mỗi một người dân yếu ớt tức làm cho cả nước yếu ớt một phần*). According to Hồ Chí Minh, a country is only so healthy as its citizens are strong (*mỗi một người dân mạnh khỏe, tức là góp phần cho cả nước mạnh khỏe*). Exercise and health are thus the "duty of every patriot" (*bổn phận của mỗi một người dân yêu nước*), he declared in his 1946 "Call for People to Exercise" (Trương 2022).

30. There were few foreigners in Hanoi during the pandemic. Most expatriates, fearing for their safety and not wanting to be separated from family members, evacuated before international borders closed on March 25. Those who remained, including individuals like me who arrived *after* March 1, were granted automatic visa extensions as a humanitarian

gesture to assist those deemed "stranded" with no means to return home. Foreigners who arrived *before* March 1 were eligible for extensions only if they could furnish embassy-issued certification that their continued stay in Vietnam was beyond their control. The rationale for choosing March 1 as the cutoff date for the humanitarian extension was not made clear.

31. The government's reluctance to report and eventalize the first death was evident in the media's response. The initial notification I received from *VnExpress* contained a link with information about the fatality, followed by other "breaking news" alerts. However, five minutes later, the "live updates" and web pages had been removed in what appeared to be a media blackout across Vietnamese news sources.

32. Returnees to Hanoi faced strict protocols: home quarantine, online medical declaration, and mandatory symptom reporting to public health authorities. Due to the large number of Hanoians who had vacationed in Đà Nẵng in July, only those returning after July 20 were tested. One interlocutor who returned on July 18 was denied a test by her local health station unless she was symptomatic, reflecting concerns about test kit shortages and overwhelmed screening facilities amid the unexpected surge in demand.

33. These predictions were later shown to be underestimated. Vietnam's GDP recorded one of the highest growth rates in the world for 2020, at 2.9 percent.

34. Vietnam was also recognized as one of ASEAN's top-performing economies during the pandemic. It stood, along with Laos, as one of the few Southeast Asian countries to achieve GDP growth in both 2020 (2.9 percent) and 2021 (2.6 percent). In 2020, Myanmar and Laos recorded higher GDP growth rates (3.2 percent and 3.3 percent, respectively). However, Myanmar experienced a crushing economic downturn of 18.6 percent in the following year after the 2021 coup, according to the Organization for Economic Cooperation and Development. DOI: https://doi.org/10.1787/e712f278-en.

35. In asking how care "figures in normative visions of a governmental good," Sylvia Tidey (2022, 91) notes that care is not exclusively a generative force for good governance but also "carries its own forms of uncertainty, unreliability, and possibilities for estrangement."

36. See, for example, Mayorga (2023), Muhammad (2019), and Johnson (2014).

37. On postcolonial ghosts, see Joseph-Vilain and Misrahi-Barak (2009).

38. Street naming practices in the Old Quarter follow the precolonial tradition of assigning names based on the craft and artisan guild associated with the production and sale of their respective goods. The original thirty-six guilds comprising the Old Quarter—or area of old streets (Khu Phố cổ)—lend their vernacular name to "36 phố phường," or the neighborhood of thirty-six streets.

39. Campt's (2017, 6) approach of listening to, rather than merely looking at, images as a way to challenge the equivalency of vision with knowledge by engaging with sound resonated with my experience with urban acoustic environments in pandemic Hanoi, despite the vastly different context of her work on rupture and refusal in photographic archives of the African Diaspora.

40. On the concept of "architectural voice" ascribed to aging buildings, see Littlefield and Lewis (2007).

41. Primarily geared toward children, the mid-autumn festivities of Tết trung thu are often described in English-language media as Halloween and Christmas rolled into one holiday.

PANDEMIC FIGURE 5: THE PHILIPPINE SINGER

1. The borders would remain closed to international air travel for two years, until March 2022. During that time, only Vietnamese citizens on embassy-organized repatriation flights, diplomats, and contracted foreign experts on government-approved flights were allowed to land in the country.

2. For many of us grappling with constrained migration and mobility in Vietnam during the pandemic, the allegations of corruption were not surprising. In my case, even securing a visa extension arranged by my neighborhood travel agent required paying an inflated fee to the Immigration Department before humanitarian exemptions were implemented. Others expressed outrage at the scale of the fraudulent operations, however. For instance, a Vietnamese student at my university managed to obtain at short notice a coveted seat on a high-demand repatriation flight by paying the booking agent in Hanoi a thousand-dollar premium over and above the government-fixed price. Considering an average of three hundred passengers per flight and twenty such repatriation flights between the United States and Vietnam in 2020 alone, this "fee bump" likely generated over six million dollars. Extrapolating globally, *VnExpress* reported that "from early 2020 to mid-2021, more than one thousand repatriation flights were organized to bring home more than two hundred thousand citizens from 62 countries and territories," hinting at colossal sums potentially accumulated through excess charges (Pham 2023).

3. According to the secretary-general of the International Maritime Organization, Kitack Lim, up to four hundred thousand seafarers were in need of repatriation as of September 2020. See www.imo.org/en/MediaCentre/PressBriefings/pages/Crew-change -COVID-19.aspx.

4. According to a September 2020 report by National Nurses United, while Filipinos (both migrants and Filipino Americans) constitute only 4 percent of registered nurses in the United States, they accounted for 31.5 percent of all COVID-related nurse mortalities, and over 50 percent of deaths among nurses of color. See www.nationalnursesunited.org /sites/default/files/nnu/documents/0920_Covid19_SinsOfOmission_Data_Report.pdf.

5. Filipino labor migrants have contributed to Vietnam's economic growth across various sectors. In 2010, while conducting fieldwork in Vinh City, Nghệ An Province, I encountered Filipino technicians at the TH True Milk eco-farm, which utilized Israeli dairy technology, New Zealand heifers, and European machinery.

6. A pseudonym to ensure anonymity.

7. Several Filipino musicians I met in Hanoi over the years of my research took up residency at the hotels where they performed nightly. Since Cecilia and her band played at different venues across the city, they were not subject to this arrangement. Thus, unlike resident musicians who could develop longer-term sonic relationships with specific spaces and loyal listeners, Cecilia and her band had to demonstrate greater sonic flexibility, continually adapting their sound to different acoustic environments and audiences.

CODA

1. See, for example, the Pandemic Sensory Archive: www.archiveofintimacy.com /sensory-archive/; the Cities and Memory #StayHomeSounds archive: https://citiesand-memory.com/covid19-sounds/; and the Library of Congress's COVID-19 oral history project: https://storycorps.org/covid-19-american-history-project/. For an analysis of how

these digital archival platforms affectively and collectively produce an intimate public, see Collins (2024).

2. Following Dylan Robinson (2020, 49–51), "guest listening" refers to an ethical practice of listening-in-relation that engages with sound through both sensory and emotional attunement. This approach "feels" history with the ears and heart, emphasizing respectful, care-based listening that honors historical and cultural resonances, in contrast to extractive "hungry" settler listening, which treats sonic practices, such as Indigenous song, as resources to be mined.

3. The extractive nature of such listening has been central to early anthropological impulses to produce audio recordings of songs and other forms of auditory expression, often in collaboration with US imperial projects to establish civilizational hierarchies. Balance (2016) illustrates this dynamic in her discussion of the Philippine reservation at the 1904 World's Fair, where the sonic figuration of people as innately primitive was mediated through recording technologies that transformed voices into artifacts.

4. This everyday fluid repertoire of anarchival traces (Massumi 2016) contrasted with the state's efforts to establish an archive of heroic sacrifice to shape future memory of the pandemic, centered around state-promoted values of care, compassion, and collective responsibility. These efforts were exemplified by visual and material artifacts, including solidarity stamps released in April 2020 (see figure 17) and a virus-themed photography book published that September titled *Tinh thần Việt và cuộc chiến chống đại dịch COVID-19* (*Vietnamese Spirit and the Fight against COVID-19*).

5. This was attributed to a relatively short national lockdown lasting only a few weeks, which maintained manufacturing productivity, minimized job losses, and sustained consumer spending.

6. A recent example of this occurred during the devastating wildfires in Maui. Some Native Hawaiians dubbed the catastrophic situation "plantation disaster capitalism" to describe settler colonial practices that dispossessed Indigenous people of their land and resources, diverting critical water supplies from Native communities (Klein and Sprout 2023).

7. For example, in 2020, the Vietnamese government evacuated 1.3 million people ahead of the massive typhoon Molave, underscoring the disproportionate impact of climate change on developing countries. At the 2021 World Climate Summit, Dr. Saleemul Huq, a climate scientist who represented Bangladesh, a nation particularly vulnerable to global warming and where socialism is enshrined in the constitution, reacted to the nearly two hundred casualties from Germany's disastrous flash flooding earlier that year, suggesting that poorer countries had valuable lessons to offer Europe on adapting to climate change: "In Bangladesh, people don't die from floods like [in Germany]. We evacuate" (Breeze 2021). Typhoon Molave, one of Vietnam's worst storms in recent history, battered the central coast, resulting in just over forty deaths.

8. Despite pronounced social stratification and a rise in the number of ultra-wealthy individuals, the Organisation for Economic Cooperation and Development (OECD) reported that in 2021, Vietnam's estimated income inequality was among the lowest in the region; www.oecd-ilibrary.org/economics/viet-nam-s-income-inequality-is-the-lowest -in-the-region_987dcd27-en. However, significant disparities persist between ethnic groups, with Kinh (Việt) and Hoa (Chinese) populations maintaining a higher socioeconomic status compared to highland minorities.

9. As a sociocultural mechanism, expectations of redistribution shape community relations and help keep inequality in check *relative to others*. While wealth is acceptable,

being excessively rich without contributing to collective well-being or flaunting wealth in ways that reinforce social disparities is frowned upon. For example, after we moved into a house in the provincial capital of Cao Bằng in 2011, our landlord, who was also the neighborhood leader (*tổ trưởng*), used our security deposit to purchase a generator. During a power outage a few days later, she illuminated her entire home, prompting a public dispute with our neighbors. After that, the generator quietly retired.

This relational thinking also shaped public responses to the COVID-19 bribery and corruption scandals that erupted in 2022–23, a sober reminder that disaster socialism remains an aspirational ideal rather than a concrete reality or guaranteed outcome. These incidents resulted in dozens of high-ranking officials being fired or imprisoned, sparking widespread anger and disillusionment among the population. One interlocutor remarked that while he had expected and even tolerated some level of corruption related to the pandemic, viewing it as intrinsic to everyday life in Vietnam, the scale of this disaster profiteering and its impact on vulnerable people rendered it unpardonable.

In both examples, individuals in positions of power—the neighborhood leader and high-ranking officials—faced ostracism and exclusion for misusing their authority and failing to exhibit self-restraint. By serving their own interests before those of the people they represented, they violated cultural expectations of redistribution and self-sacrifice for the collective good.

10. I left Vietnam in November 2020 for Germany, staying there until vaccines became available in the United States. When I applied to reenter Vietnam in June 2021, my return was delayed by the Delta surge, which had restricted entry for experts and family members alike. After four months navigating bureaucratic paperwork—including new burdensome regulations for legalizing documents like marriage and CDC vaccine certificates—I finally received the necessary approvals from multiple ministries and government agencies. Upon my return, I found a city transformed by color-coded risk zones that determined which businesses could operate and which remained shuttered.

11. Loudspeaker broadcasts in the time of Delta called for city leaders to "mobilize the entire political system and all sectors of society to participate in pandemic prevention efforts . . . [while] upholding a sense of responsibility" (*huy động cả hệ thống chính trị và các tầng lớp nhân dân tham gia phòng chống dịch . . . nêu cao tinh thần trách nhiệm*).

12. See, for example, Le (2022). The image accompanying this article shows a warden coordinating vaccinations in his neighborhood of nine hundred residents, using a megaphone to instruct people on how to prepare for their appointments.

13. The most publicized incident, captured on video and widely criticized by Vietnamese netizens and the media, involved a young worker at a checkpoint in Nha Trang. Though he carried proper documentation and was permitted to travel to secure essential food items, he was stopped for purchasing a baguette. A patrolling officer confiscated the worker's motorbike and, pointing a finger in his face, used a derogatory form of address to harshly declare, "Bread is not essential food" (Bánh mì không phải thực phẩm thiết yếu). This phrase quickly went viral, inciting mass outrage online. Public reaction and media attention led to disciplinary measures against the offending officers for their unprofessional conduct and prompted authorities to clarify the definition of "essential food." For a sharp critique of the "poor quality" (*chất lượng yếu kém*) of grassroots leadership during the pandemic that this incident exposed, see Tường Minh (2021). My thanks to Sarah Grant for bringing this incident to my attention.

14. The survey can be accessed at www.undp.org/vietnam/publications/citizens-opinion-and-experiences-government-responses-covid-19-pandemic-viet-nam-2nd-round-survey-2021. On the 2020 survey, see act 2, note 39.

15. Critics have urged me to think more deeply about "party-state brutality" and the "heavily militarized" lockdown of this period. While I appreciate this push, this view does not reflect how my interlocutors in Hồ Chí Minh City experienced the role of the army, which was twofold: delivering food provisions and monitoring established checkpoints. When critique was directed, it was more often aimed at individuals who abused their power, such as provincial leaders or police (see above), rather than at the presence of young soldiers. Contrary to assumptions of a detached and uncaring state, many residents I spoke with viewed the military as collaborating with citizens to make a brutally long—but to many necessary—lockdown survivable under extremely difficult circumstances. They struggled with my questions and skepticism about the military's role, perceiving me as either uninformed or influenced by Western perspectives that did not align with their lived experience of the crisis.

It is important to note that the soldiers were regarded as *essential workers* undertaking tasks for which they had not been trained, such as shopping and delivering food. The state-promoted concept of "*tình quân dân*" (military-civilian bond), which has its roots in Vietnam's revolutionary history, shaped public perceptions, even in a historically complex place like former Saigon. My interlocutors tended to express gratitude towards the young conscripts for their work, which included emotionally charged tasks, such as delivering cremated remains to families, one person recounted to me through tears. Online comics and graphics humorously yet critically depicted the logistical challenges faced by these unarmed (and often rural) forces, highlighting their struggles to navigate unfamiliar urban environments, locate obscure items on shopping lists, and accommodate diverse needs, hinting at underlying class differences. Graphic images that circulated of young, and often confused, soldiers laden with colorful shopping bags served to humanize their presence, casting them not as oppressive figures but as victims of an unmanageable situation themselves. While this perspective may diverge from more critical geopolitical narratives about Vietnam's response to the pandemic, it reconfirms the findings of other scholars studying neighborhood networks during the crisis. See Tough (2021), for instance, who expresses skepticism not about the role of the military but about the effectiveness of some of its actions, like mass disinfection of urban streets. Thank you to Quynh Truong for their social media work on the comics.

16. In Vietnamese: "Đẩy mạnh công tác tuyên truyền hiệu quả hơn, nhanh hơn bằng nhiều hình thức thiết thực, phát huy tuyên truyền trên loa, truyền thanh, xã phường, các khu chung cư, trên không gian mạng, các hình thức tuyên truyền lưu động."

17. Twelve thousand kilograms is equivalent to 26.5 thousand pounds.

18. For more on the Muzik Dập Dịch campaign, see https://baodantoc.vn/muzik-dap-dich-gop-nhac-thanh-gao-1629426406288.htm.

Acland, Charles R., ed. 2007. *Residual Media*. Minneapolis: University of Minnesota Press.

Adams, Vincanne. 2013. *Markets of Sorrow, Labors of Faith: New Orleans in the Wake of Katrina*. Durham, NC: Duke University Press.

———. 2020. "Disasters and Capitalism . . . and COVID-19." *Somatosphere*, March 26. http://somatosphere.net/2020/disaster-capitalism-covid19.html/.

Agamben, Giorgio. 2005. *State of Exception*. Translated by Kevin Attell. Chicago: University of Chicago Press.

———. 2021. *Where Are We Now?: The Epidemic as Politics*. New York: Rowman and Littlefield.

Ahuja, Neel. 2016. *Bioinsecurities: Disease Interventions, Empire, and the Government of Species*. Durham, NC: Duke University Press.

Allison, Anne. 2013. *Precarious Japan*. Durham, NC: Duke University Press.

Altés Arlandis, Alberto, and Oren Lieberman. 2021. "Intraventions in Flux: Towards a Modal Spatial Practice that Moves and Cares." In *Architecture and Collective Life*, edited by Penny Lewis, Lorens Holm, and Sandra Costa Santos, 251–59. London: Routledge.

Anderson, Ben. 2016. "Emergency/Everyday." In *Time: A Vocabulary of the Present*, edited by Joel Burges and Amy J. Elias, 177–91. New York: New York University Press.

Anderson, Warwick. 2006. *Colonial Pathologies: American Tropical Medicine, Race, and Hygiene in the Philippines*. Durham, NC: Duke University Press.

Antona, Laura. 2020. "The New Normal or Same Old? The Impacts of the Covid-19 Pandemic on Live-in Migrant Domestic Workers in Singapore." *LSE Southeast Asia Blog*, November 4. https://blogs.lse.ac.uk/seac/2020/11/04/the-new-normal-or-same-old-the-impacts-of-the-covid-19-pandemic-on-live-in-migrant-domestic-workers-in-singapore/.

Appadurai, Arjun. 2020. "The COVID Exception." *Social Anthropology* 28 (2): 221–22.

Aso, Michitake. 2013. "Patriotic Hygiene: Tracing New Places of Knowledge Production about Malaria in Vietnam, 1919–75." *Journal of Southeast Asian Studies* 44 (3): 423–43.

Atkinson, Rowland. 2007. "Ecology of Sound: The Sonic Order of Urban Space." *Urban Studies* 44 (10): 1905–17.

Attali, Jacques. (1985) 2009. *Noise: The Political Economy of Music*. Translated by Brian Massumi. Minneapolis: University of Minnesota Press.

Balance, Christine Bacareza. 2016. *Tropical Renditions: Making Musical Scenes in Filipino America*. Durham, NC: Duke University Press.

Barad, Karen. 2007. *Meeting the Universe Halfway: Quantum Physics and the Entanglement of Matter and Meaning*. Durham, NC: Duke University Press.

Barker, Joshua, Erik Harms, and Johan Lindquist. 2014. *Figures of Southeast Asian Modernity*. Honolulu: University of Hawai'i Press.

Barthes, Roland. 1972. *Mythologies*. Translated by Jonathan Cape. London: Farrar, Straus & Giroux.

Bayly, Susan. 2020. "Beyond 'Propaganda': Images and the Moral Citizen in Late-Socialist Vietnam." *Modern Asian Studies* 54 (5): 1526–95.

Becquelin, Aurore Monod, and Jacques Galinier, eds. 2020. "And Then Came the Night…: Field Work, Methods, Perspectives." *Ateliers d'Anthropologie* 48. DOI: https://doi.org/10.4000/ateliers.13380.

Belsie, Laurent. 2021. "From Lumber to Labor, Are We Now in a 'Shortage Economy'?" *Christian Science Monitor*, May 11. https://www.csmonitor.com/USA/Politics/2021/0511/From-lumber-to-labor-are-we-now-in-a-shortage-economy.

Benjamin, Walter. 1969. "Theses on the Philosophy of History." In *Illuminations*, edited by Hannah Arendt, translated by H. Zohn, 253–64. New York: Schocken.

Berlant, Lauren. 2011. *Cruel Optimism*. Durham, NC: Duke University Press.

Biehl, João. 2016. "Theorizing Global Health." *Medicine Anthropology Theory* 3 (2): 127–42.

Birdsall, Carolyn. 2012. *Nazi Soundscapes: Sound, Technology and Urban Space in Germany, 1933–1945*. Amsterdam: University of Amsterdam Press.

Bolis, Mara, Anam Parvez, Emma Holten, Leah Mugehera, Nabil Abdo, and Maria Jose Moreno. 2020. "Care in the Time of Coronavirus: Why Care Work Needs to Be at the Centre of a Post-COVID-19 Feminist Future." Oxfam. DOI: 10.21201/2020.6232. https://oxfamilibrary.openrepository.com/bitstream/handle/10546/621009/bp-care-crisis-time-for-global-reevaluation-care-250620-en.pdf.

Bolter, Jay David, and Richard Arthur Grusin. 2000. *Remediation: Understanding New Media*. Cambridge, MA: MIT Press.

Brakman, Sarah-Vaughan. 2020. "Social Distancing Isn't a Personal Choice. It's an Ethical Duty." *Vox*, April 9. https://www.vox.com/future-perfect/2020/4/9/21213425/coronavirus-covid-19-social-distancing-solidarity-ethics.

Brecht, Bertolt. (1932) 1967. "Der Rundfunk als Kommunikationsapparat" [The radio as apparatus of communication]. In *Schriften zur Literatur und Kunst I (1920–1932)*, 132–40. Frankfurt am Main: Suhrkamp.

Breeze, Nick. 2021. "Dr Saleemul Huq Interview." *ClimateGenn Podcast*, November 26. https://genn.cc/dr-saleemul-huq-interview-vulnerable-countries-left-glasgow-with-tears-in-their-eyes/.

Brooks, Daphne A. 2006. *Bodies in Dissent: Spectacular Performances of Race and Freedom, 1850–1910*. Durham, NC: Duke University Press.

———. 2016. "How #BlackLivesMatter Started a Musical Revolution." *Guardian*, March 13. https://www.theguardian.com/us-news/2016/mar/13/black-lives-matter-beyonce-kendrick-lamar-protest.

Brotherton, P. Sean. 2012. *Revolutionary Medicine: Health and the Body in Post-Soviet Cuba.* Durham, NC: Duke University Press.

Brown, Nik, Sarah Nettleton, Chrissy Buse, Alan Lewis, and Daryl Martin. 2021. "The Coughing Body: Etiquettes, Techniques, Sonographies and Spaces." *Biosocieties* 16 (2): 270–88.

Bui, Long. 2020. "Coronavirus and America's Vietnam Syndrome." *Orange County Register*, April 30. https://www.ocregister.com/2020/04/30/coronavirus-and-americas-vietnam-syndrome/.

Butler, Judith. 1990. *Gender Trouble.* New York: Routledge.

Cadière, Léopold. 1910. "Sur quelques faits religieux ou magiques observés pendant une épidémie de choléra en Annam" [On some religious or magical practices observed during a cholera epidemic in Annam]. *Anthropos* 5 (2): 1125–59.

Cadogan, Garnette. 2016. "Walking while Black." *Literary Hub*, July 8. https://lithub.com/walking-while-black/.

Caduff, Carlo. 2014. "Sick Weather Ahead: On Data-Mining, Crowd-Sourcing and White Noise." *Cambridge Journal of Anthropology* 32 (1): 32–46.

Caldwell, Melissa L. 2009. "Introduction: Food and Everyday Life after State Socialism." In *Food and Everyday Life in the Postsocialist World*, edited by Melissa L. Caldwell, 1–28. Bloomington: Indiana University Press.

Campt, Tina M. 2017. *Listening to Images.* Durham, NC: Duke University Press.

Cardoso, Leonardo. 2018. "Sound-Politics in São Paulo: Noise Control and Administrative Flows." *Current Anthropology* 59 (2): 192–208.

Care Collective. 2020. *The Care Manifesto: The Politics of Interdependence.* London: Verso.

Cassiman, Ann, Thomas H. Eriksen, and Lotte Meinert. 2022. "Pandemic Precarity." *Anthropology Today* 38 (4): 1–2.

Chadwick, Andrew. 2017. *The Hybrid Media System: Politics and Power.* 2nd ed. Oxford: Oxford University Press.

Charles, Nicole. 2022. *Vaccines, Hesitancy, and the Affective Politics of Protection in Barbados.* Durham, NC: Duke University Press.

Chen, Mel Y. 2012. *Animacies: Biopolitics, Racial Mattering, and Queer Affect.* Durham, NC: Duke University Press.

Chen, Nancy, and Lesley Sharp, eds. 2014. *Bioinsecurity and Vulnerability.* Santa Fe, NM: School for Advanced Research Press.

Chin, Elizabeth. 2016. *My Life with Things: The Consumer Diaries.* Durham, NC: Duke University Press.

Choi, Vivian Y. 2015. "Anticipatory States: Tsunami, War, and Insecurity in Sri Lanka." *Cultural Anthropology* 30 (2): 286–309.

Choy, Catherine Ceniza. 2003. *Empire of Care: Nursing and Migration in Filipino American History.* Durham, NC: Duke University Press.

Chua, Lynette J., and Jack Jin Gary Lee. 2021. "Governing through Contagion." In *COVID-19 in Asia: Law and Policy Contexts*, edited by Victor V. Ramraj, 115–32. New York: Oxford.

Classen, Constance. 1997. "Foundations for an Anthropology of the Senses." *International Social Sciences Journal* 153: 401–12.

Coderre, Laurence. 2021. *Newborn Socialist Things: Materiality in Maoist China.* Durham, NC: Duke University Press.

Collier, Stephen J., and Andrew Lakoff. 2008. "The Problem of Securing Health." In *Biosecurity Interventions: Global Health and Security in Question*, edited by Andrew Lakoff and Stephen J. Collier, 7–32. New York: Columbia University Press.

Collins, Emily 2024. "Listening Together/Apart: Intimacy and Affective World-Building in Pandemic Digital Archival Sound Projects." *Sounding Out!*, January 22. https://soundstudiesblog.com/2024/01/22/listening-together-apart-intimacy-and-affective-world-building-in-pandemic-digital-archival-sound-projects/.

Corbin, Alain. 1986. *The Foul and the Fragrant: Odor and the French Social Imagination.* Cambridge, MA: Harvard University Press.

———. 1998. *Village Bells: Sound and Meaning in the 19th Century French Countryside.* New York: Columbia University Press.

Corwin, Julia E., and Vinay Gidwani. 2021. "Repair Work as Care: On Maintaining the Planet in the Capitalocene." *Antipode.* DOI: https://doi.org/10.1111/anti.12791.

Covid Crisis Group. 2023. *Lessons from the Covid War: An Investigative Report.* New York: PublicAffairs.

Craig, David. 2002. *Familiar Medicine: Everyday Health Knowledge and Practice in Today's Vietnam.* Honolulu: University of Hawai'i Press.

Dao, Amy. 2024. "Trust in Numbers? The Politics of Zero Deaths and Vietnam's Response to COVID-19. In *Reconfiguring Global Societies in the Pre-Vaccination Phase of the COVID-19 Pandemic,* edited by Jack Fong, 134–54. Toronto: University of Toronto Press.

Das, Veena. 1995. *Critical Events: An Anthropological Perspective on Contemporary India.* Oxford: Oxford University Press.

Daughtry, J. Martin. 2015. *Listening to War: Sound, Music, Trauma, and Survival in Wartime Iraq.* New York: Oxford University Press.

de Certeau, Michel. 1984. *The Practice of Everyday Life.* Translated by Steven Rendall. Berkeley: University of California Press.

Derryberry, Elizabeth P., Jennifer N. Phillips, Graham E. Derryberry, Michael J. Blum, and David Luther. 2020. "Singing in a Silent Spring: Birds Respond to a Half-Century Soundscape Reversion during the COVID-19 Shutdown." *Science* 370 (6516): 575–79. DOI: https://doi.org/10.1126/science.abd5777.

Dewachi, Omar. 2017. *Ungovernable Life: Mandatory Medicine and Statecraft in Iraq.* Palo Alto, CA: Stanford University Press.

Dunn, Elizabeth C., and Katherine Verdery. 2011. "Dead Ends in the Critique of (Post)Socialist Anthropology: Reply to Thelen." *Critique of Anthropology* 31 (3): 251–55.

Ebbighausen, Rodion. 2020. "How Vietnam Is Winning Its 'War' on Coronavirus." *Deutsche Welle,* April 16. https://www.dw.com/en/how-vietnam-is-winning-its-war-on-coronavirus/a-52929967.

Edwards, Claire, and Nicola Maxwell. 2023. "Troubling Ambulant Research: Disabled People's Socio-Spatial Encounters with Urban Un/Safety and the Politics of Mobile Methods." *Irish Journal of Sociology* 31 (1): 63–81.

Eidsheim, Nina Sun. 2019. *The Race of Sound: Listening, Timbre, and Vocality in African American Music.* Durham, NC: Duke University Press.

Elkin, Lauren. 2016. *Flâneuse: Women Walk the City in Paris, New York, Tokyo, Venice, and London.* New York: Farrar, Straus and Giroux

Engelmann, Lukas. 2020. "#COVID19: The Spectacle of Real-Time Surveillance." *Somatosphere,* March 6. http://somatosphere.net/forumpost/covid19-spectacle-surveillance/.

Erlmann, Veit. 2004. "But What of the Ethnographic Ear? Anthropology, Sound and the Senses." In *Hearing Cultures: Essays in Sound, Listening, and Modernity,* edited by Veit Erlmann, 1–20. New York: Berg.

Fahlman, Miles Joseph. 2019. "The Anthropology of SARS and the Leveraging of Cultural Logics in Vietnam." Master's thesis, University of Saskatchewan.

Fairhead, James. 2016. "Understanding Social Resistance to the Ebola Response in the Forest Region of the Republic of Guinea: An Anthropological Perspective." *African Studies Review* 59 (3): 7–31.

Fanon, Frantz. 1965. *A Dying Colonialism*. Translated by Haakon Chevalier. New York: Grove.

Farina, Almo. 2014. *Soundscape Ecology: Principles, Patterns, Methods and Applications*. New York: Springer.

Fearnley, Lyle. 2008. "Signals Come and Go: Syndromic Surveillance and Styles of Biosecurity." *Environment and Planning A: Economy and Space* 40 (7): 1615–32.

Fehérváry, Krisztina. 2009. "Goods and States: The Political Logic of State-Socialist Material Culture." *Comparative Studies in Society and History* 51 (2): 426–59.

Feld, Steven. 1996. "Waterfalls of Song: An Acoustemology of Place Resounding in Bosavi, Papua New Guinea." In *Senses of Place*, edited by Steven Feld and Keith Basso, 91–135. Santa Fe, NM: School of American Research Press.

———. 2015. "Acoustemology." In *Keywords in Sound*, edited by David Novak and Matt Sakakeeny, 12–21. Durham, NC: Duke University Press.

Fernandes, Sujatha. 2006. *Cuba Represent!: Cuban Arts, State Power, and the Making of New Revolutionary Cultures*. Durham, NC: Duke University Press.

Ferrarini, Lorenzo, and Nicola Scaldaferri. 2020. *Sonic Ethnography: Identity, Heritage and Creative Research Practice in Basilicata, Southern Italy*. Manchester: Manchester University Press.

Fisher, Marc, Scott Wilson, and Arelis R. Hernández. 2020. "Covid's U.S. Toll: Nearly 300,000 Dead and a Stalemate Between Fatalism and Hope." *Washington Post*, December 12. https://www.washingtonpost.com/national/covid-300000-dead-us/2020/12/12/5f40686c-364f-11eb-8d38-6aea1adb3839_story.html.

Foster, Hal. 2004. "An Archival Impulse." *October* 110: 3–22.

Foucault, Michel. 1973. *The Birth of the Clinic: An Archaeology of Medical Perception*. Translated by A. M. Sheridan. New York: Random House.

———. 1977. *Discipline and Punish: The Birth of the Prison*. Translated by Alan Sheridan. New York: Vintage Books.

———. 1978. *The History of Sexuality*. Vol. 1, *An Introduction*. Translated by Robert Hurley. New York: Random House.

———. 1982. "The Subject and Power." *Critical Inquiry* 8 (4): 777–95.

———. 1991. "Question of Method." In *The Foucault Effect: Studies in Governmentality*, edited by Graham Burchell, Colin Gordon, and Peter Miller, 73–86. Chicago: University of Chicago Press.

Gabauer, Angelika, Sabine Knierbein, Nir Cohen, Henrik Lebuhn, Kim Trogal, Tihomir Viderman, and Tigran Haas, eds. 2021. *Care and the City: Encounters with Urban Studies*. London: Routledge.

Galinier, Jacques, Aurore Monod Becquelin, Guy Bordin, Laurent Fontaine, Francine Fourmaux, Juliette Roullet Ponce, Piero Salzarulo, Philippe Simonnot, Michèle Therrien, and Iole Zilli. 2010. "Anthropology of the Night: Cross-Disciplinary Investigations." *Current Anthropology* 51 (6): 819–47.

Gamble, Julie. 2017. "Experimental Infrastructure: Experiences in Bicycling in Quito, Ecuador." *International Journal of Urban and Regional Research* 41 (1): 162–80.

Gammeltoft, Tine M. 2007. "Prenatal Diagnosis in Postwar Vietnam: Power, Subjectivity, and Citizenship." *American Anthropologist* 109 (1): 153–63.

———. 2014. *Haunting Images: A Cultural Account of Selective Reproduction in Vietnam*. Berkeley: University of California Press.

———. 2022. "Everyday Attentiveness: Understanding Diabetes in Vietnam through Literary Displacement." *Journal of the Royal Anthropological Institute* 28 (2): 595–612.

Gelles, David. 2021. "When Disaster Hits Home for a Global Aid Organization." *New York Times*, January 8. https://www.nytimes.com/2021/01/08/business/raj-shah-rockefeller-foundation-corner-office.html.

Giles, David Boarder. 2021. *A Mass Conspiracy to Feed People*. Durham, NC: Duke University Press.

Glickstein, Jonathan A. 2002. *American Exceptionalism, American Anxiety: Wage, Competition, and Degraded Labor in the Antebellum United States*. Charlottesville: University of Virginia Press.

Goffman, Erving. (1956) 1990. *The Presentation of Self in Everyday Life*. London: Penguin.

Grondin, Simon, Esteban Mendoza-Duran, and Pier-Alexandre Rioux. 2020. "Pandemic, Quarantine, and Psychological Time." *Frontiers in Psychology* 11. DOI: https://doi.org/10.3389/fpsyg.2020.581036.

Grossman, Zoltán. 2021. "The Resilience Doctrine: An Introduction to Disaster Resilience." *Counterpunch*, February 1. https://www.counterpunch.org/2021/02/01/the-resilience-doctrine-an-introduction-to-disaster-resilience/.

Gui, Weihsin. 2024. "Covidity and Comics: Graphic Medicine from Singapore. *ImageText* 15 (1). https://imagetextjournal.com/gui-covidity/.

Gupta, Sujata. 2022. "How Living in a Pandemic Distorts our Sense of Time." *Science News*, September 12. https://www.sciencenews.org/article/pandemic-living-distorts-time-sense.

Guyton, Shelley T. 2022. "National Disaster Imaginary: Mediatized Disaster and Filipino Subjectivity." In *Environment, Media, and Popular Culture in Southeast Asia*, edited by Jason Paolo Telles, John Charles Ryan, and Jeconiah Louis Dreisbach, 291–304. Singapore: Springer.

Guzman, Elena H., and Emily Hong. 2022. "Feminist Sensory Ethnography: Embodied Filmmaking as a Politic of Necessity." *Visual Anthropology Review* 38 (2): 184–210.

Ha An. 2020. "Food-Hailing Action Nothing but a Drop in the Ocean of Plastic." *Vietnam Investment Review*, August 24–30, 24.

Hage, Ghassan. 2009. "Waiting Out the Crisis: On Stuckedness and Governmentality." In *Waiting*, edited by Ghassan Hage, 97–106. Carlton: Melbourne University Press.

———. 2015. *Alter-Politics Critical Anthropology, Political Passion and the Radical Imagination*. Melbourne: Melbourne University Press.

Haider, Najmul, Alexei Yavlinsky, Yu-Mei Chang, Mohammad N. Hasan, Camilla Benfield, Abdinasir Y. Osman, MD Jamal Uddin, Osman Dar, Francine Ntoumi, Alimuddin Zumla, and Richard Kock. 2020. "The Global Health Security Index and Joint External Evaluation Score for Health Preparedness Are Not Correlated with Countries' COVID-19 Detection Response Time and Mortality Outcome." *Epidemiology & Infection* 148 (E210). DOI: https://doi.org/10.1017/S0950268820002046.

Halstead, Narmala. 2020. "Exceptions and Being Human: Before and in Times of COVID-19." *Journal of Legal Anthropology* 4 (1): v–xii. DOI: https://doi.org/10.3167/jla.2020.040101.

Hansen, Arve. 2017. "Hanoi on Wheels: Emerging Automobility in the Land of the Motorbike." *Mobilities* 12 (5): 628–45.

Hayton, Bill, and Tro Ly Ngheo. 2020. "Vietnam's Coronavirus Success Is Built on Repression." *Foreign Policy*, May 12. https://foreignpolicy.com/2020/05/12/vietnam-coronavirus-pandemic-success-repression/.

Helmreich, Stefan. 2007. "An Anthropologist Underwater: Immersive Soundscapes, Submarine Cyborgs, and Transductive Ethnography." *American Ethnologist* 34 (4): 621–41.

———. 2016. *Sounding the Limits of Life: Essays in the Anthropology of Biology and Beyond.* Princeton, NJ: Princeton University Press.

Henriques, Julian. 2011. *Sonic Bodies: Reggae Sound Systems, Performance Techniques, and Ways of Knowing.* New York: Continuum.

Herrero, Ana Vanessa, and Austin Ramzy. 2019. "Maduro Sounds Conciliatory but Warns: US Intervention Would Be Worse than Vietnam." *New York Times*, January 30. https://www.nytimes.com/2019/01/30/world/americas/maduro-venezuela-talks-opposition.html.

Hirschkind, Charles. 2006. *The Ethical Soundscape: Cassette Sermons and Islamic Counterpublics.* New York: Columbia University Press.

Hoang, Lien. 2021. "Asia's COVID Recovery: Vietnam's Breakout Moment." *Nikkei Asia*, January 20. https://asia.nikkei.com/Spotlight/The-Big-Story/Asia-s-COVID-recovery-Vietnam-s-breakout-moment.

Hobart, Hi'ilei Julia Kawehipuaakahaopulani, and Tamara Kneese. 2020. "Radical Care: Survival Strategies for Uncertain Times." *Social Text* 38 (1): 1–16.

Hồng Vân. 2020. "Công nhân Việt ở Guinea Xích đạo, Uzbekistan: Mong từng ngày được về nước" [Vietnamese workers in equatorial Guinea, Uzbekistan: Everyday hoping to return home]. *Tuổi Trẻ*, July 27. https://tuoitre.vn/cong-nhan-viet-o-guinea-xich-dao-uzbekistan-mong-tung-ngay-duoc-ve-nuoc-2020072709011363.htm.

hooks, bell. 1984. *Feminist Theory: From Margin to Center.* New York: South End Press.

Horst, Heather and Daniel Miller. 2006. *The Cell Phone: An Anthropology of Communication.* New York: Taylor & Francis.

Hosek, Jennifer Ruth. 2024. "Vaccine Patriotism and Public Health Cultures: Cuba's Scalable Best Practices in the Covid-19 Pandemic." *Canadian Journal of Latin American and Caribbean Studies* 49 (2): 190–219.

Howes, David. 2019. "Multisensory Anthropology." *Annual Review of Anthropology* 48 (1): 17–28.

———. 2022. "In Defense of Materiality: Attending to the Sensori-Social Life of Things." *Journal of Material Culture* 27 (3): 313–35.

Hsieh, Jennifer C. 2021. "Making Noise in Urban Taiwan: Decibels, the State, and Sono-Sociality." *American Ethnologist* 48 (1): 51–64.

Hsu, Elisabeth. 2008. "The Senses and the Social: An Introduction." *Ethnos* 73 (4): 433–43.

Hull, Matthew. 2012. *Government of Paper: The Materiality of Bureaucracy in Urban Pakistan.* Berkeley: University of California Press.

Huynh, Toan Luu Duc. 2020. "The COVID-19 Containment in Vietnam: What are We Doing?" *Journal of Global Health* 10 (1). DOI: https://doi.org/10.7189/jogh.10.010338.

Icaza, Rosalba. 2017. "Decolonial Feminism and Global Politics: Border Thinking and Vulnerability as a Knowing Otherwise." In *Critical Epistemology of Global Politics*, edited by Marc Woons and Sebastian Weier, 26–45. Bristol: E-International Relations.

Ingold, Tim. 2000. *The Perception of the Environment: Essays on Livelihood, Dwelling and Skill*. New York: Routledge.

———. 2004. "Culture on the Ground: The World Perceived through the Feet." *Journal of Material Culture* 9 (3): 315–40.

———. 2010. "Footprints through the Weather-World: Walking, Breathing, Knowing." *Journal of the Royal Anthropological Institute* 16 (S1): S121–39.

———. 2011. *Being Alive: Essays on Movement, Knowledge and Description*. New York: Routledge.

International Science Council. 2020. "The Sound of Science: The SARS-CoV-2 Virus as a Piece of Classical Music." April 19. https://council.science/current/news/the-sound-of -science-the-sars-cov-2-virus-as-a-piece-of-classical-music/.

Ivic, Sanja. 2020. "Vietnam's Response to the COVID-19 Outbreak." *Asian Bioethics Review* 12: 341–47.

Jacobs, Jane. 1961. *The Death and Life of Great American Cities*. New York: Vintage.

Janus, Adrienne. 2011. "Listening: Jean-Luc Nancy and the 'Anti-Ocular' Turn in Continental Philosophy and Critical Theory." *Comparative Literature* 63 (2): 182–202.

Johnson, Andrew Alan. 2014. *Ghosts of the New City: Spirits, Urbanity, and the Ruins of Progress in Chiang Mai*. Honolulu: University of Hawai'i Press.

Jones, Carla. 2020. "Introduction: Mourning a Moment: Affect and Crisis." In *Pandemic Diaries: Affect and Crisis*, edited by Carla Jones. American Ethnologist website. https:// americanethnologist.org/features/pandemic-diaries/pandemic-diaries-affect-and -crisis/introduction-mourning-a-moment-affect-and-crisis.

Joseph-Vilain, Mélanie, and Judith Misrahi-Barak, eds. 2009. *Postcolonial Ghosts*. Montpellier: Presses universitaires de la Méditerranée. DOI: https://doi.org/10.4000/books.pulm.10813.

Kenner, Alison. 2021. "Emplaced Care and Atmospheric Politics in Unbreathable Worlds." *Environment and Planning C: Politics and Space* 39 (6): 1113–28.

Khan, Naveeda. 2011. "The Acoustics of Muslim Striving: Loudspeaker Use in Ritual Practice in Pakistan." *Comparative Studies in Society and History* 53 (3): 571–94.

Kheshti, Roshanak. 2015. *Modernity's Ear: Listening to Race and Gender in World Music*. New York: New York University Press.

Kim, Annette. 2015. *Sidewalk City: Remapping Public Space in Ho Chi Minh City*. Chicago: Chicago University Press.

Kim, Joey S. 2020. "Orientalism in the Age of COVID-19." *Los Angeles Review of Books*, March 24. https://lareviewofbooks.org/short-takes/orientalism-age-covid-19/.

Klein, Christina. 2003. *Cold War Orientalism: Asia in the Middlebrow Imagination, 1945–1961*. Berkeley: University of California Press.

Klein, Naomi. 2007. *Shock Doctrine: The Rise of Disaster Capitalism*. New York: Penguin.

Klein, Naomi, and Kapua'ala Sproat. 2023. "Why Was There No Water to Fight the Fire in Maui?" *Guardian*, August 17. https://www.theguardian.com/commentisfree/2023 /aug/17/hawaii-fires-maui-water-rights-disaster-capitalism.

Kleinman, Arthur, and Sing Lee. 2005. "SARS and the Problem of Social Stigma." In *SARS in China: Prelude to Pandemic?*, edited by Arthur Kleinman and James L. Watson, 173–95. Palo Alto, CA: Stanford University Press.

Klinenberg, Eric. 2002. *Heat Wave: A Social Autopsy of Disaster in Chicago*. Chicago: University of Chicago Press.

Koh, David W. H. 2006. *Wards of Hanoi*. Singapore: ISEAS.

Kornai, János. 1980. *Economics of Shortage*. Vol. A. New York: North-Holland.

Kotsko, Adam. 2022. "What Happened to Giorgio Agamben?" *Slate*, February 20. https://slate.com/human-interest/2022/02/giorgio-agamben-covid-holocaust-comparison-right-wing-protest.html.

LaBelle, Brandon. 2018. *Sonic Agency: Sound and Emergent Forms of Resistance*. London: Goldsmiths Press.

Lacey, Kate. 2013. *Listening Publics: The Politics and Experience of Listening in the Media Age*. New York: Polity.

Lakoff, Andrew. 2017. *Unprepared: Global Health in a Time of Emergency*. Berkeley: University of California Press.

Larkin, Brian. 2004. "Degraded Images, Distorted Sounds: Nigerian Video and the Infrastructure of Piracy." *Public Culture* 16 (2): 289–314.

———. 2014. "Techniques of Inattention: The Mediality of Loudspeakers in Nigeria." *Anthropological Quarterly* 87 (4): 989–1015.

Latour, Bruno. 2020. "Is This a Dress Rehearsal?" *Critical Inquiry Online Blog*, March 26. https://critinq.wordpress.com/2020/03/26/is-this-a-dress-rehearsal/.

Le, Lam. 2022. "Vietnam Vaccinates 90% of its Population against COVID-19." *Gavi*, February 28. https://www.gavi.org/vaccineswork/vietnam-vaccinates-90-its-population-against-covid-19.

Lê Thanh Tâm and Lưu Ngọc. 2022. "Loa phường: bỏ hay giữ?" [Ward loudspeakers: Abolish or maintain?] *Tuổi Trẻ*, July 28. https://tuoitre.vn/loa-phuong-bo-hay-giu-20220728090239079.htm.

Lee, Marie Myung-Ok. 2020. "'Wuhan Coronavirus' and the Racist Art of Naming a Virus." *Salon*, February 6. https://www.salon.com/2020/02/06/the-racist-art-of-naming-a-virus/.

Lefebvre, Henri. 2013. *Rhythmanalysis: Space, Time and Everyday Life*. Translated by Stuart Elden and Gerard Moore. New York: Continuum.

Liboiron, Max. 2021. *Pollution Is Colonialism*. Durham, NC: Duke University Press.

Lincoln, Martha. 2014. "Tainted Commons, Public Health: The Politico-Moral Significance of Cholera in Vietnam." *Medical Anthropology* Quarterly 28 (3): 342–61.

———. 2021. *Epidemic Politics in Contemporary Vietnam: Public Health and the State*. London: Bloomsbury Academic.

Lippman, Alexandra. 2021. "Sonic Governance: Culturalization and Criminalization of *Funk Carioca* in Rio de Janeiro." *Anthropological Quarterly* 94 (3): 443–72.

Littlefield, David, and Saskia Lewis. 2007. *Architectural Voices: Listening to Old Buildings*. Hoboken, NJ: Wiley

Liu, Chen, Trung Thang Nguyen, and Yujiro Ishimura. 2021. "Current Situation and Key Challenges on the Use of Single-Use Plastic in Hanoi." *Waste Management* 121: 422–31. DOI: https://doi.org/10.1016/j.wasman.2020.12.033.

Logan, Amanda L. 2020. *The Scarcity Slot: Excavating Histories of Food Security in Ghana*. Berkeley: University of California Press.

Lorenzini, Daniele. 2021. "Biopolitics in the Time of Coronavirus." *Critical Inquiry* 47 (2): S40–S45. DOI: https://doi.org/10.1086/711432.

Loria, Jeffrey H., and Julie Loria, eds. 2021. *Silent Cities: Portraits of a Pandemic: 15 Cities Across the World*. New York: Skyhorse.

Lovell, Stephen. 2015. *Russia in the Microphone Age: A History of Soviet Radio, 1919–1970*. Oxford: Oxford University Press.

Low, Kelvin E. Y. 2015. "The Sensuous City: Sensory Methodologies in Urban Ethnographic Research." *Ethnography* 16 (3): 295–312.

———. 2023. *Sensory Anthropology: Culture and Experience in Asia*. Cambridge. UK: Cambridge University Press.

Low, Kelvin E. Y., and Noorman Abdullah. 2021. "Sensory Experience as Method." In *The Oxford Handbook of the Sociology of the Body and Embodiment*, edited by Natalie Boero and Katherine Mason, 53–68. Oxford: Oxford University Press.

Lynteris, Christos, and Branwyn Poleykett. 2018. "The Anthropology of Epidemic Control: Technologies and Materialities." *Medical Anthropology* 37 (6): 433–41.

Maierhofer-Lischka, Margarethe. 2024. "About Loudspeakers and Hearing in Times of Change." *Sounding Futures*. https://www.soundingfuture.com/en/article/about-loudspeakers-and-hearing-times-change.

Malarney, Shaun Kingsley. 2011. Introduction to *Lục Xì: Prostitution and Venereal Disease in Colonial Hanoi*, by Vũ, Trọng Phụng, 1–44. Honolulu: University of Hawai'i Press.

———. 2012. "Germ Theory, Hygiene and the Transcendence of 'Backwardness' in Revolutionary Vietnam (1954–60)." In *Southern Medicine for Southern People: Vietnamese Medicine in the Making*, edited by Laurence Monnais, C. Michele Thompson and Ayo Wahlberg, 107–32. Newcastle upon Tyne: Cambridge Scholars.

Malmström, Frederika, M. 2022. "Making and Unmaking Masculinities in Cairo through Sonic Infrastructural Violence." *Urban Studies* 59 (3): 591–607.

Marr, David. 1981. *Vietnamese Tradition on Trial, 1920–1945*. Berkeley: University of California Press.

Martin, Allie. 2019. "Hearing Change in the Chocolate City: Soundwalking as Black Feminist Method." *Sounding Out!*, August 5. https://soundstudiesblog.com/2019/08/05/hearing-change-in-the-chocolate-city-soundwalking-as-black-feminist-method/.

Massumi, Brian. 2016. "Working Principles." In *The Go-to How to Book of Anarchiving*, edited by Andrew Murphie, 6–7. Montreal: Sense Lab.

Mattern, Shannon. 2015. "Deep Time of Media Infrastructure." *Signal Traffic: Critical Studies of Media Infrastructures*, edited by Lisa Parks and Nicole Starosielski, 94–112. Champaign: University of Illinois Press.

———. 2020. "Purity and Security: Towards a Cultural History of Plexiglass." *Places Journal*. DOI: https://doi.org/10.22269/201201.

Mattingly, Cheryl. 2019. "Waiting." *Cambridge Journal of Anthropology* 37 (1): 17–31.

Max D. T. 2020. "The Public Shaming Pandemic." *New Yorker*, September 21. https://www.newyorker.com/magazine/2020/09/28/the-public-shaming-pandemic.

Mayorga, Sarah. 2023. *Urban Specters: The Everyday Harms of Racial Capitalism*. Chapel Hill, NC: University of North Carolina Press.

McKittrick, Katherine. 2011. "On Plantations, Prisons, and a Black Sense of Place." *Social & Cultural Geography* 12 (8): 947–63.

Meikle, Graham, and Sherman Young. 2012. *Media Convergence: Networked Digital Media in Everyday Life*. London: Palgrave Macmillan.

Mitchell, W. J. T. 2004. *What Do Pictures Want? The Lives and Loves of Images*. Chicago: University of Chicago Press.

Mol, Annemarie. 2008. *The Logic of Care: Health and the Logic of Patient Choice*. New York: Routledge.

Monnais, Laurence. 2006. "Preventative Medicine and 'Mission Civilisatrice': Uses of the BCG Vaccine in French Colonial Vietnam Between the Two World Wars." *International Journal of Asia Pacific Studies* 2 (1): 40–66.

Muhammad, Ismail. 2019. "Walking with the Ghosts of Black Los Angeles." *Literary Hub*, September 20. https://lithub.com/walking-with-the-ghosts-of-black-los-angeles/.

Murphy, Michelle. 2015. "Unsettling Care: Troubling Transnational Itineraries of Care in Feminist Health Practices." *Social Studies of Science* 45 (5): 717–37.

———. 2017. *Economization of Life*. Durham, NC: Duke University Press.

Mydans, Seth. 2002. "Americans and Vietnamese Fighting over Catfish." *New York Times*, November 5. https://www.nytimes.com/2002/11/05/world/americans-and-vietnamese-fighting-over-catfish.html.

Nakano, Takashi and Tomoya Onishi. 2020. "Vietnam Emerges as Sole Economic Winner in Southeast Asia." *Nikkei Asia*, November 19. https://asia.nikkei.com/Economy/Vietnam-emerges-as-sole-economic-winner-in-Southeast-Asia.

Nancy, Jean-Luc. 2007. *Listening*. Translated by Charlotte Mandel. New York: Fordham University Press.

Navaro-Yashin, Yael. 2009. "Affective Spaces, Melancholic Objects: Ruination and the Production of Anthropological Knowledge." *Journal of the Royal Anthropological Institute* 15 (1): 1–18.

Ng, Stephanie. 2005. "Performing the 'Filipino' at the Crossroads: Filipino Bands in Five-Star Hotels throughout Asia." *Modern Drama* 48 (2): 272–96.

Nguyen, Hong Kong, and Tung Manh Ho. 2020. "Vietnam's COVID-19 Strategy: Mobilizing Public Compliance via Accurate and Credible Communications." *Perspective (ISEAS)* 69: 1–15.

Nguyen, Long. 2020. "Urban Exodus a New Trend among Young Vietnamese," *VnExpress*, October 20. https://e.vnexpress.net/news/life/trend/urban-exodus-a-new-trend-among-young-vietnamese-4175501.html.

Nguyen, Minh T. N. 2016. "Trading in Broken Things: Gendered Performances and Spatial Practices in a Northern Vietnamese Rural-Urban Waste Economy." *American Ethnologist* 43 (1): 116–29.

Nguyen, Sen. 2020. "Vietnam's Pandemic Success Is a Lesson for the World." *Global Asia* 15 (3). https://www.globalasia.org/v15no3/cover/vietnams-pandemic-success-is-a-lesson-for-the-world_sen-nguyen.

Nguyen, Thi Nguyet Minh. 2010. "Migrants and Non-migrant Domestic Workers in Hanoi: The Segmentation of Domestic Service." Working Paper 22, DEV Working Paper Series, School of International Development, University of East Anglia, UK.

Nguyen Cong Khai, Pham Quang Thai, Ha-Linh Quach, et al. 2020. "Transmission of SARS-CoV 2 during Long-Haul Flight." *Emerging Infectious Diseases* 26 (11): 2617–24.

Nguyễn Thị Phương Châm. 2022. "Hà Nội Sidewalks: Mediating Urban Order in Common Spaces." *Journal of Vietnamese Studies* 17 (4): 1–17.

Nguyễn Trung Kiên. 2022. "Chủ tịch Hồ Chí Minh với việc phát triển các môn thể thao truyền thống" [President Hồ Chí Minh and the development of traditional sports]. *Tuyên Giáo*, March 30. https://bvhttdl.gov.vn/chu-tich-ho-chi-minh-voi-viec-phat-trien-cac-mon-the-thao-truyen-thong-20220517094649733 htm

Nguyen-Thu Giang. 2020. "From Wartime Loudspeakers to Digital Networks: Communist Persuasion and Pandemic Politics in Vietnam." *Media International Australia* 177 (1): 144–48.

Ó Briain, Lonán. 2022. *Voices of Vietnam: A Century of Radio, Red Music, and Revolution.* Oxford: Oxford University Press.

Ochoa Gautier, Ana María. 2014. *Aurality: Listening and Knowledge in Nineteenth-Century Colombia.* Durham, NC: Duke University Press.

Older, Malka. 2016. "Securitization of Disaster Response in the United States." *Revue Interdiciplinaire de Travaux sur les Amériques* [Interdisciplinary review of work on the Americas]. https://hal.archives-ouvertes.fr/hal-01520627/.

Oliveros, Pauline. 2011. "Auralizing in the Sonosphere: A Vocabulary for Inner Sound and Sounding." *Journal of Visual Culture* 10 (2): 162–68.

Oliver-Smith, Anthony. 1999. "The Brotherhood of Pain: Theoretical and Applied Perspectives on Post-Disaster Solidarity." In *The Angry Earth: Disaster in Anthropological Perspective,* edited by Anthony Oliver-Smith and Susannah Hoffman, 156–72. New York: Routledge.

Östersjö, Stefan, and Nguyễn Thanh Thủy. 2016. "The Sounds of Hanoi and the After-Image of the Homeland." *Journal of Sonic Studies* 12. https://www.researchcatalogue.net/view/246523/246546.

Panimbang, Fahmi. 2021. "Solidarity across Boundaries: A New Practice of Collectivity among Workers in the App-Based Transport Sector in Indonesia." *Globalizations* 18 (8): 1377–91.

Parreñas, Rhacel Salazar. 2001. *Servants of Globalization: Women, Migration, and Domestic Work.* Palo Alto, CA: Stanford University Press.

Perry, Brea L., Brian Aronson, and Bernice A. Pescosolido. 2021. "Pandemic Precarity: COVID-19 is Exposing and Exacerbating Inequalities in the American Heartland." *Proceedings of the National Academy of Sciences* 118 (8). DOI: https://doi.org/10.1073/pnas.2020685118.

Peterson, Marina. 2021. *Atmospheric Noise: The Indefinite Urbanism of Los Angeles.* Durham, NC: Duke University Press.

Pham Du. 2023. "How Briberies Launched Vietnam's Pandemic Repatriation Flights." *VnExpress,* April 6. https://e.vnexpress.net/news/news/how-briberies-launched-vietnam-s-pandemic-repatriation-flights-4590340.html.

Phạm Duy. 2020. "Chôn chân giữa 'biển người' đi chơi Trung thu trên phố Hàng Mã" [Buried in a 'sea of people' celebrating Mid-Autumn Festival on Hàng Mã Street]. *Tiền Phong,* September 29. https://tienphong.vn/chon-chan-giua-bien-nguoi-di-choi-trung-thu-tren-pho-hang-ma-post1278133.tpo.

Phạm, Linh. 2020. "A Personal History of Hồ Tây: Romance, Colonial Rule, and Subsidy-Era Fishing Heists." *Saigoneer,* March 25. https://saigoneer.com/hanoi-heritage/24593-a-personal-history-of-hồ-tây-romance,-colonial-rule-and-subsidy-era-fishing-heists.

Phan Huy Lê. 2018. "Đền Cầu Nhi di tích độc đáo giữa Hà thành" [Cầu Nhi Temple: A unique relic in the middle of the city]. *Tài Nguyên và Môi Trường,* February 6. https://baotainguyenmoitruong.vn/den-cau-nhi-di-tich-doc-dao-giua-ha-thanh-286829.html.

Phương Nhung 2020. "Lãnh đạo phường Trúc Bạch nhảy 'Ghen cô Vy' sau khi dỡ lệnh cách ly," [Trúc Bạch leaders dance to "Ghen cô Vy" after lifting the quarantine]. *Dân Trí,* March 23. https://dantri.com.vn/doi-song/lanh-dao-phuong-truc-bach-nhay-ghen-co-vy-sau-khi-do-lenh-cach-ly-20200323074105779.htm.

Pijanowski, Bryan C., Luis J. Villanueva-Rivera, Sarah L. Dumyahn, Almo Farina, Bernie L. Krause, Brian M. Napoletano, Stuart H. Gage, and Nadia Pieretti. 2011. "Soundscape Ecology: The Science of Sound in the Landscape." *BioScience* 61 (3): 203–16.

Pink, Sarah. 2011. "Multimodality, Multisensoriality and Ethnographic Knowing: Social Semiotics and the Phenomenology of Perception." *Qualitative Research* 11 (3): 261–76.

———. 2015. *Doing Sensory Ethnography*. 2nd ed. London: Sage.

Porath, Nathan, ed. 2019. *Hearing Southeast Asia: Sounds of Hierarchy and Power in Context*. Copenhagen: NIAS Press.

Porter, Natalie. 2019. *Viral Economies: Bird Flu Experiments in Vietnam*. Chicago: University of Chicago Press.

Prats, Armando José. 2002. *Invisible Natives: Myth and Identity in the American Western*. Ithaca, NY: Cornell University Press.

Proschan, Frank. 2002. "'Syphilis, Opiomania, and Pederasty': Colonial Constructions of Vietnamese (and French) Social Diseases." *Journal of the History of Sexuality* 11 (4): 610–36.

Przybylski, Liz. 2021. *Hybrid Ethnography: Online, Offline, and in Between*. London: Sage.

Puig de la Bellacasa, Maria. 2011. "Matters of Care in Technoscience: Assembling Neglected Things." *Social Studies of Science* 41 (1): 85–106.

Quỳnh Hoa. 2020. "Loa phường 'hồi sinh' thời dịch bệnh" [Pandemic "resurrects" ward loudspeakers]. *Sức khỏe và Đời sống*, April 27. https://suckhoedoisong.vn/loa-phuong -hoi-sinh-thoi-dich-benh-169173113.htm.

Rachwał, Tadeusz. 2017. *Precarity and Loss: On Certain and Uncertain Properties of Life and Work*. Berlin: Springer.

Rebuscini, Afra, and Yuri Frassi. 2022. *Crafting a Sonic City*. Officine Gặp online publication. https://freight.cargo.site/m/Y1892971199342852329499745165797/Crafting-a-sonic-city .pdf.

Remnick, David. 2023. "The Pandemic at Three: Who Got It Right," *New Yorker Radio Hour*, February 24. https://www.newyorker.com/podcast/the-new-yorker-radio-hour/the -pandemic-at-three-who-got-it-right.

Rice, Tom. 2003. "Soundselves: An Acoustemology of Sound and Self in the Edinburgh Royal Infirmary." *Anthropology Today* 19 (4): 4–9.

———. 2013. *Hearing and the Hospital: Sound, Listening, Knowledge and Experience*. Hereford, UK: Sean Kingston.

Robert, Christoph. 2020. "Visualizing the Invisible: COVID-19 Pandemic Season in Saigon." *EchoGéo* 52. DOI: https://doi.org/10.4000/echogeo.19692.

Robinson, Dylan. 2020. *Hungry Listening: Resonant Theory for Indigenous Sound Studies*. Minneapolis: University of Minnesota Press.

Rogaski, Ruth. 2004. *Hygienic Modernity: Meanings of Health and Disease in Treaty-Port China*. Berkeley: University of California Press.

Rose-Redwood, Reuben, Natchee Blu Barnd, Annita Hetoevėhotohkeʼe Lucchesi, Sharon Dias, and Wil Patrick. 2020. "Decolonizing the Map: Recentering Indigenous Mappings." *Cartographica* 55 (3): 151–62.

Roth, Emmanuelle. 2020. "Epidemic Temporalities: A Concise Literature Review." *Anthropology Today* 36 (4): 13–16.

Saguin, Kristian Karlo. 2022. *Urban Ecologies on the Edge: Making Manila's Resource Frontier*. Berkeley: University of California Press.

Sahlins, Marshall. 1985. *Islands of History*. Chicago: Chicago University Press.

Saji, Sweetha, Sathyaraj Venkatesan, and Brian Callender. 2021. "Comics in the Time of a Pan(dem)ic: COVID-19, Graphic Medicine, and Metaphors." *Perspectives in Biology and Medicine* 64 (1): 136–54.

Samuels, David W., Louise Meintjes, Ana Maria Ochoa, and Thomas Porcello. 2010. "Soundscapes: Toward a Sounded Anthropology." *Annual Review of Anthropology* 39 (1): 329–45.

Sangaramoorthy, Thurka, and Adia Benton. 2020. "Imagining Rural Immunity." *Anthropology News*, June 19. DOI: 10.1111/AN.1439. https://www.anthropology-news.org/articles/imagining-rural-immunity/?fbclid=IwAR0STo8D7ZIvgrMo3S4mekie7zTI4A3qFR0fAV4F5Y5VvEZlkEn9jFONBvw.

Scarry, Elaine. 2011. *Thinking in an Emergency*. New York: W. W. Norton.

Schafer, R. Murray. 1994. *The Soundscape: Our Sonic Environment and the Tuning of the World*. Rochester, VT: Destiny.

Schuurman, Nadine, and Geraldine Pratt. 2002. "Care of the Subject: Feminism and Critiques of GIS." *Gender, Place & Culture* 9 (3): 291–99. DOI: https://doi.org/10.1080/0966369022000003905.

Schwenkel, Christina. 2009a. *The American War in Contemporary Vietnam: Transnational Remembrance and Representation*. Bloomington: Indiana University Press.

———. 2009b. "'The Camera Was My Weapon': News Production and Representation of War in Vietnam." In *The Anthropology of News and Journalism: Global Perspectives*, edited by S. Elizabeth Bird, 86–99. Bloomington: Indiana University Press.

———. 2019. "Governing through Garbage: Waste Infrastructure Breakdown and Gendered Apathy in Vietnam." In *Routledge Handbook of Anthropology and the City*, edited by Setha Low, 318–31. New York: Routledge.

———. 2020. *Building Socialism: The Afterlife of East German Architecture in Urban Vietnam*. Durham, NC: Duke University Press.

———. 2022a. "The Things They Carried (and Kept): Revisiting 'Ostalgie' in the Global South." *Comparative Studies in Society and History* 64 (2): 478–509.

———. 2022b. "Social Housing and Feminist Commoning in Urban Vietnam." *Focaal: Journal of Global and Historical Anthropology* 94: 20–37. DOI: https://doi.org/10.3167/fcl.2022.940102.

Schwenkel, Christina, and Ann Marie Leshkowich. 2012. "How Is Neoliberalism Good to Think Vietnam? How Is Vietnam Good to Think Neoliberalism?" *positions: asia critique* 20 (2): 379–401.

Scoones, Ian, Rebecca Smalley, Ruth Hall, and Dzodzi Tsikata. 2019. "Narratives of Scarcity: Framing the Global Land Rush." *Geoforum* 101: 231–41.

Scott, Sarah Mayberry. 2021. "Sonic Lessons of the Covid-19 Soundscape." *Sounding Out!*, August 2. https://soundstudiesblog.com/2021/08/02/sonic-lessons-of-the-covid-19-soundscape/.

Shah, Sonia. 2020. "It's Time to Tell a New Story about the Coronavirus—Our Lives Depend on It." *Nation*, July 14. https://www.thenation.com/article/society/pandemic-definition-covid/.

Sharma, Sarah. 2014. *In the Meantime: Temporality and Cultural Politics*. Durham, NC: Duke University Press.

Shohet, Merav. 2021. *Silence and Sacrifice: Family Stories of Care and the Limits of Love in Vietnam*. Berkeley: University of California Press.

Siisiäinen, Lauri. 2013. *Foucault and the Politics of Hearing*. New York: Routledge.

Sims, Holly, and Kevin Vogelmann. 2002. "Popular Mobilization and Disaster Management in Cuba." *Public Administration and Development* 22 (5): 389–400.

Sitrin, Marina, and Colectiva Sembrar, eds. 2020. *Pandemic Solidarity: Mutual Aid during the Covid-19 Crisis*. London: Pluto Press.

Small, Christopher. 1998. *Musicking: The Meanings of Performing and Listening*. Middletown, CT: Wesleyan University Press.

Solnit, Rebecca. 2001. *Wanderlust: A History of Walking*. New York: Penguin.

———. 2009. *A Paradise Built in Hell: The Extraordinary Communities that Arise in Disaster*. New York: Penguin.

Sontag, Susan. 1978. *Illness as Metaphor*. New York: Farrar, Straus & Giroux.

Sorokin, Vladimir. (1983) 2008. *The Queue*. Translated by Sally Laird. New York: New York Review Books Classics.

Springgay, Stephanie, and Sarah E. Truman. 2019. *Walking Methodologies in a More-than-Human World: WalkingLab*. New York: Routledge.

Srnicek, Nick. 2017. *Platform Capitalism*. Malden, MA: Polity.

Staffa, Rachel K., Maraja Riechers, and Berta Martín-López. 2022. "A Feminist Ethos for Caring Knowledge Production in Transdisciplinary Sustainability Science." *Sustainability Science* 17: 45–63.

Sterne, Jonathan. 2003. *The Audible Past: Cultural Origins of Sound Reproduction*. Durham, NC: Duke University Press.

Stewart, Kathleen. 2011. "Atmospheric Attunements." *Environment and Planning D: Society and Space* 29: 445–53.

Stoever, Jennifer Lynn. 2016. *The Sonic Color Line: Race and the Cultural Politics of Listening*. New York: New York University Press.

Stoler, Ann Laura. 2004. "Affective States." In *A Companion to the Anthropology of Politics*, edited by David Nugent and Joan Vincent, 4–20. New York: Blackwell.

Sun Media. 2020. "Cruelty in Reports Black Cats Killed for COVID-19 Cure." *Toronto Sun*, April 24. https://torontosun.com/news/world/cruelty-in-reports-black-cats-killed-for-covid-19-cure.

Tai, Hue-Tam Ho. 1992. *Radicalism and the Origins of the Vietnamese Revolution*. Cambridge, MA: Harvard University Press.

Tausig, Benjamin. 2019. *Bangkok is Ringing: Sound, Protest, and Constraint*. Oxford: Oxford University Press.

Thái Bá Dũng. 2021. "Tổ trưởng ơi . . ." [Hey neighborhood leader . . .]. *Tuổi trẻ*, August 17. https://tuoitre.vn/to-truong-oi-20210816231630764.htm.

Thompson, Emily. 2002. *The Soundscape of Modernity: Architectural Acoustics and the Culture of Listening in America, 1900–1933*. Cambridge, MA: MIT Press.

Tidey, Sylvia. 2022. *Ethics or the Right Thing? Corruption and Care in the Age of Good Governance*. Chicago: HAU.

Tough, Rachel. 2021. "Ho Chi Minh City During the Fourth Wave of COVID-19 in Vietnam." *City & Society* 33 (3). DOI: https://doi.org/10.1111/ciso.12413.

Tran, Allen L. 2023. *A Life of Worry: Politics, Mental Health, and Vietnam's Age of Anxiety*. Berkeley: University of California Press.

Trần Lê Hưng. 2021. "Xây dựng hình mẫu con người mới xã hội chủ nghĩa" [Building a new model of socialist persons]. *Đảng Cộng sản Việt Nam*, September 4.

Trần Mạnh and Đình Nam. 2020. "'Cuộc chiến' chống Covid-19: Bắt đầu chiến dịch mới" [The "battle" against COVID-19: The start of a new campaign]. *Hà Nội Mới*, March 8. http://hanoimoi.com.vn/tin-tuc/suc-khoe/960538/cuoc-chien-chong-covid-19-bat-dau-chien-dich-moi.

Trnka, Susanna. 2020. "Rethinking States of Emergency." *Social Anthropology/Anthropologie sociale* 28 (2): 367–68. DOI: https://doi.org/10.1111/1469-8676.12812.

Trnka, Susanna, and Catherine Trundle. 2014. "Competing Responsibilities: Moving beyond Neoliberal Responsibilisation." *Anthropological Forum* 24 (2): 136–53.

Tronto, Jane C. 2013. *Caring Democracy: Markets, Equality, and Justice*. New York: New York University Press.

Truitt, Allison. 2013. *Dreaming of Money in Ho Chi Minh City*. Seattle: University of Washington Press.

Trương Quốc Uyên. 2022. *Chủ tịch Hồ Chí Minh với thể dục thể thao* [President Hồ Chí Minh and sports]. Hà Nội: Thể dục Thể thao.

Tường Minh. 2021. "Bánh mì không phải là thực phẩm: Một bộ phận cán bộ cơ sở quá yếu kém" [Bread is not food: Some grassroots cadres lack capacity]. *Lao Động*, July 20. https://laodong.vn/ban-doc/banh-mi-khong-phai-la-thuc-pham-mot-bo-phan-can-bo-co-so-qua-yeu-kem-932471.ldo.

Turner, Sarah, and Ngô Thúy Hạnh. 2019. "Contesting Socialist State Visions for Modern Mobilities: Informal Motorbike Taxi Drivers' Struggles and Strategies on Hanoi's Streets, Vietnam." *International Development Planning Review* 41 (1): 43–61.

Turner, Victor. 1967. *The Forest of Symbols: Aspects of Ndembu Ritual*. Ithaca, NY: Cornell University Press.

Uddin, Mahi. 2021. "Addressing Work-Life Balance Challenges of Working Women during COVID-19 in Bangladesh." *International Social Science Journal* 71 (239–40): 7–20.

Vallas, Steven, and Juliet B. Schor. 2020. "What Do Platforms Do? Understanding the Gig Economy." *Annual Review of Sociology* 46 (1): 273–94.

Vann, Michael G., and Liz Clarke. 2018. *Great Hanoi Rat Hunt: Empire, Disease, and Modernity in French Colonial Vietnam*. Oxford: Oxford University Press.

Verdery, Katherine. 1996. *What Was Socialism, and What Comes Next?* Princeton, NJ: Princeton University Press.

Việt Nam News. 2020. "The Truth about COVID-19 Patient 17 in Việt Nam." September 28. https://vietnamnews.vn/opinion/op-ed/772833/the-truth-about-covid-19-patient-17-in-viet-nam.html.

Viết Tuân. 2020. "Cách ly toàn xã hội không phải là phong tỏa đất nước" [Total society quarantine is not a national lockdown]. *VnExpress*, March 31. https://vnexpress.net/cach-ly-toan-xa-hoi-khong-phai-la-phong-toa-dat-nuoc-4077606.html.

Vo Hai. 2017. "War-Time Loudspeakers to Continue Blaring out across Hanoi Despite Huge Public Opposition." *VnExpress*, April 1. https://e.vnexpress.net/news/news/war-time-loudspeakers-to-continue-blaring-out-across-hanoi-despite-huge-public-opposition-3564352.html.

Vo, Hoang-Long, Hoai A. S. Nguyen, Khanh Ngoc Nguyen, et al. 2020. "Adherence to Social Distancing Measures for Controlling COVID-19 Pandemic: Successful Lesson from Vietnam." *Frontiers in Public Health* 8. DOI: https://doi.org/10.3389/fpubh.2020.589900.

Voegelin, Salomé. 2010. *Listening to Noise and Silence: Towards a Philosophy of Sound Art*. New York: Continuum.

————. 2021. *Sonic Possible Worlds: Hearing the Continuum of Sound*. Rev. ed. New York: Bloomsbury Academic.

von Busekist, Astrid. 2001. "L'indicible" [The unutterable]. *Raisons politiques* 2 (2): 89–112.

Vũ Nguyên. 2020. "Netizen Việt giận cá chém thớt vụ nam tiếp viên 'bệnh nhân 1342'" [Vietnamese netizens lash out over the male flight attendant "Patient 1342"]. *Tuổi Trẻ*, December 3. https://cuoi.tuoitre.vn/netizen-viet-gian-ca-chem-thot-vu-nam-tiep-vien -benh-nhan-1342-20201203472159999.htm.

Vũ Trọng Phụng. 2011. *Lục Xì: Prostitution and Venereal Disease in Colonial Hanoi*. Translated by Shaun Kingsley Malarney. Honolulu: University of Hawai'i Press.

Wald, Priscilla. 2008. *Contagious: Cultures, Carriers, and the Outbreak Narrative*. Durham, NC: Duke University Press.

Watanabe, Chika. 2021. "Playing through the Apocalypse: Preparing Children for Mass Disasters in Japan and Chile." *Public Culture* 33 (2): 239–59.

Weheliye, Alexander G. 2005. *Phonographies: Grooves in Sonic Afro-Modernity*. Durham, NC: Duke University Press.

Welna, David. 2020. "Coronavirus Has Now Killed More Americans than Vietnam War." *NPR*, April 28. https://www.npr.org/sections/coronavirus-live-updates/2020/04/28/846701304 /pandemic-death-toll-in-u-s-now-exceeds-vietnam-wars-u-s-fatalities.

Westerkamp, Hildegard. 2017. "The Practice of Listening in Unsettled Times." Keynote, "Invisible Places: Sound Urbanism and Sense of Place," April 7, São Miguel Island, Azores, Portugal. https://www.hildegardwesterkamp.ca/writings/writingsby/?post_id= 61&title=practice-of-listening-in-unsettled-times.

Williams, Alex. 2020. "The Year of the Blur." *New York Times*, October 31. https://www .nytimes.com/2020/10/31/style/the-year-of-blur.html.

Wilson, Elizabeth. 1992. "The Invisible Flâneur." *New Left Review* 1 (191): 90–110.

Wisner, B. G. 1978. "Letter to the Editor." *Disasters* 2 (1): 80–82.

Wunderlich, Filipa Matos. 2008. "Walking and Rhythm City: Sensing Urban Space." *Journal of Urban Design* 13 (1): 125–39.

X. M. 2020. "Người dân khu vực cách ly tại phố Trúc Bạch 'chống dịch COVID-19' bằng luyện tập thể dục thể thao" [Quarantined residents in Trúc Bạch "fight COVID-19" by exercising]. *Báo Tin tức*, March 9. https://baotintuc.vn/xa-hoi/nguoi-dan-khu-vuc -cach-ly-tai-pho-truc-bach-chong-dich-covid19-bang-luyen-tap-the-duc-the-thao -20200308184735316.htm.

yamomo, meLê. 2018. *Theatre and Music in Manila and the Asia Pacific, 1869–1946*. Cham: Palgrave Macmillan.

Young, Alison. 2021. "The Limits of the City: Atmospheres of Lockdown." *British Journal of Criminology* 61 (4): 985–1004.

accountability, 10, 30
active listening, 54–56
Adele, 135
ADP. *See* Asian Development Bank
affective labor, 134–35
Agamben, Giorgio, 155n1
Agent Orange, 45
air pollution, 90
alternative sonic life-worlds, 86, 92
ambient sounds, 67–68
Anderson, Ben, 111
anthrophonics, 118–25
anthropological research, during COVID-19
        pandemic, 13–14
anthropomorphism, xiv
anticipatory listening, 54–55
anticolonial struggles, 74–75, 124–25
anti-communism, 10, 32
Appadurai, Arjun, 11
Architectural Voices (Littlefield and
        Lewis), 125
Asian Development Bank (ADP), 127
Astra (COVID-19 vaccine), 145
ATMs, rice and mask, 45, 143
Attali, Jacques, 84, 86
audile techniques, 139
auditory cues. *See* sonic cues
authority, vocalizations of, 63
autoethnography, 14
avian flu, 35; in Vietnam, 9–10

Ba Đình District, Hanoi, 1, 7, 39
Bắc Ninh Province, 95
Bạch Mai Hospital, 63, 159n29
bảng thông báo (notice board), 80, 163n34
baristas, 84
Bayly, Susan, 16
bia hơi lunch gatherings, 88–89
bicycles, 47, 100, 167n6
biodata, 49
biodefense, 36
biopower, 12, 16, 21
biosecurity, 32, 47, 125; in Global North, 11–12;
        screening to maintain, 26, 38
birds, 118; birdsong and, 114–16
Blackness, sound and, 15, 88, 109; sonic dissent
        and, 165n47; walking and, 106–7
Bluezone (app), 33
Blursday, 69
Brecht, Bertolt, 70, 75
Brotherton, Sean, 16–17
Buddha Bar, 63, 160n3
building industry, 90–91
Busekist, Astrid von, 89–90

Cadogan, Garnette, 106
Caldwell, Melissa, 156n12
California, 40–41
cardio beats, 123–25
care: e-care, 32, 56; emphasis on collective
        forms of, 35, 103, 124, 132; ethics of, 14–15;

care *(continued)*
packages, 43–44; in quarantine, 48–56; reciprocal care relations, 111*fig.*; relational approach to, 16, 20, 50 147*fig.*; state care and control, 22, 55; urban caretaking, 39–40, 58, 59; in urban environments, 127; care workers, 23, 50, 58, 61, 81. *See also* sonic governance of care

Cầu Nhi Temple, 169n21

Charles, Nicole, 87

Châu Long wet market, 42, 48

Chế Lan Viên, 99

chiến sĩ áo trắng, 25, 77

Chin, Elizabeth, 14

Chinese Cultural Revolution, xvi

citizenship: caring and, 16; hygienic, 18; moral, 77; urban, 87

civil liberties, 85–86

Coderre, Laurence, xvi

cohabitation, 99, 131

cohort quarantines, 5, 32

Cold War, 32; experiences of scarcity during, 42; Orientalism in, 10

collectivism, disaster, 5

colonialism, 10; French colonialism, 18–19, 73, 143, 152n6; settler colonialism, 107, 173n6

Communiqué 1560/BYT-MT, 51

compliance, 32; performance of, 84–87. *See also* noncompliance

Confucianism, 85, 152n3

consent, 32, 153n16

construction, sound of, 90–91

contact tracing, 63, 151; apps, 33, 99–100

cooperation, 32, 105

Corbin, Alain, 69

coughing: sensory attunement to, 79; sonic cues and, 15, 26, 66

counterpublic, 62–63

COVAX, 144, 152n11

COVID-19 pandemic, xiii, xv; affectsphere of, xvi–xvii; anthropological research during, 13–14; anticipatory management of, 63; consumer goods shortages during, 40, 40*fig.*; corruption scandal and, 139, 172n2; deaths from, 130, 141–42, 144; global responses to, 9–13; initial outbreaks in, xx–xxi; militarized rhetoric and, 131, 144, 151n4, 152n9, 152n13; Orientalism and, 3; power dynamics in, 32; preparedness for, 36–37; public health infrastructure breakdown during, xvii–xviii; public responses to restrictions during, 33; sensory dimensions of, 7; social types in, 23; sonic campaigns of, 77–82; sound in, 17; temporality of, 69; United States' response to, xviii, 11, 142; vaccines, 145, 157n19, 174n10; Vietnamese response to, xviii, xx, 141–42. *See also* lockdown, COVID-19

critical cartography, 112–14

Cuba, 11, 36, 146; health care in, 16

Cúc Phương National Park, 134

đá cầu (foot badminton), 115, 169n17

Đà Nẵng, 80, 96–97, 171n32; outbreak in, 126

dangerous intimacy, 124

Đào Vân Anh, 5

Daughtry, J. Martin, 78

Decision 749/QD-TTg, 35, 156n5

decontamination, 51, 58; in Trúc Bạch Ward, 53*fig.*, 54. *See also* disinfection

Decree on Home Quarantine, 49

Defense Ministry, Vietnam, 51

deficit thinking, 10

Delta variant, COVID-19, 23, 143–44; lockdown, 86

"Đêm tàn" (Brutal Night), 99

Department of Information and Communications, 80, 162n23

Department of Preventative Medicine, 131

Điện Biên Phủ, 143

digitalization, 35–37, 72; in sonic socialism, 82; in Vietnam, 39, 156n5

Directive 15/CT-TTg, 64

Directive 16/CT-TTg, 63, 64, 94

disaster capitalism, 20, 142

disaster collectivism, 5

disaster governance: state power and, 8; Vietnamese preparedness and, 35–36

disaster socialism, 22, 23, 36, 64, 85, 105; anticipatory governance and, 142

Disease Prevention pamphlet, 49

disinfection, 51, 104, 145, 159n30

Dispatch 16212/BYT-MT, 159n30

dissent, sonic, 87–88, 91*fig.*; noise as, 91–92

distortion, 63, 138–43, 146

dogs, 67, 101*fig.*, 120

domestic workers, 23, 57–61

Douglas, Mary, 51

edging processes, 111–12

e-governance, 14, 35, 38, 71

Eidsheim, Nina Sun, 121

emergency conditions, 110–11; preparedness and, 11, 20, 85; state power and, 32

emoticons, 34

Equatorial Guinea, 133

essential workers, 23, 24–28, 43, 175n15
ethical consumption, 42
eventalization, xix, 126, 171n31
the everyday, 110–12
exceptionalism, Vietnamese, 85
exercise groups, 124*fig.*, 168n7. *See also* fitness
expatriates, 58–61, 161n13, 170n30

Fos, 5, 164n35, 167n4
Facebook, 39
fake news, 39, 79, 104
Fatherland Front, 146
Fehérváry, Krisztina, 41
feminism, 14, 153n16; soundwalking as Black
    feminist practice, 109
fence-breaking, 87, 91–92
fieldwork: proxy, 13–14; as soundwork, 13–15
fish, 117–18, 169n18; in "catfish war," 152n9;
    weaponized use of, 10
fitness, 100, 102*fig.*; groups, 123–25; revolution
    and, 125
folk songs, xv–xvi, 19, 124
foot badminton (*đá cầu*), 115, 169n17
footwork, 66, 107
Foucault, Michel, xix, 12, 16, 32, 78
Foucault and the Politics of Hearing (Siisiäinen),
    15–16
Fourth Industrial Revolution, 35, 155n5
frog market (*chợ cóc*), 118–19

Gammeltoft, Tine M., 35
the gaze, 15–16
gender: gendered soundscapes, 119, 123*fig.*, 170n25;
    inequalities and, 57; loudspeaker broadcasting
    and, 77, 162n27; musical labor and, 135
"Ghen Cô Vy" (COVID Envy), xiii–xiv, xv*fig.*,
    xix, xx, 6, 17, 30, 103, 149n3
#GhenCoVyChallenge, xiv
ghost cities, 125–32
Global North: biosecurity in, 11–12;
    donor-recipient hierarchy in, 42
God of Earth (Ông Địa), 132
Goffman, Erving, 86
Gold Week campaign, 157n20
GrabFood, 39
GrabTaxi, drivers, 94–97
grassroots communication, 81
guest listening, 138, 173n2

Hai Bà Trưng District, 81
hand sanitizers, 44*fig.*; availability of, 43;
    customer care and, 96; shortage in United
    States of, 40

handwashing dance, xiv, 5–6. *See also*
    "Ghen Cô Vy"
Hàng Mã (Votive Paper) Street, 128, 130
headphones, 162n26
health e-declarations, 31*fig. See also* medical
    declarations
Henriques, Julian, xvi, 150n5
Hồ Chí Minh, 125, 157n20
Hồ Chí Minh City, 45–46, 63, 96, 130, 144
Hồ Chí Minh Communist Youth League, 64
Hoàn Kiếm Lake, 129
Hoàng Sa Island, 65*fig.*
hospitality industry, 125–26, 134
household labor, division of, 57, 82
housekeepers (*người giúp việc*), 57–61
human-nonhuman encounters, xiv, 101*fig.*, 113;
    walking and, 102, 109, 114–18
hungry listening, 138–39
Hurricane Katrina, 36
hybrid ethnography, 13–14
hygiene, 50, 103–4, 154n28; education and, 18–20;
    hygiene nationalism, 19; public health
    campaigns and, xiii, 20, 79, 82, 103–4

identity: listening practices linked to, 81–82;
    loa phường and, 81–82; media and, 80;
    performance of, 86
Immigration Department, 172n2
imperialism, 10, 134; sound and, 18, 151n1
improvised recordings, 122
Independence Day, Vietnamese, 125
Industry 4.0, 35, 39
inequalities, gendered, 57; pandemic-related, 60,
    128, 174n9
Ingold, Tim, 169n13
instant messaging, 32, 34, 50, 54–55; COVID-19
    management via, 30–31
institutional failure, 12
interactivity, sonic ecology of, 110–11
Iraq War, 78–79
Italy, xviii, 64

Khắc Hưng, 6, 149n1
Kim Cuc, 73
Klein, Naomi, 142

lakeshore anthrophonies, 118–25
Lào Cai Province, 161n17
Larkin, Brian, 63, 139
Latour, Bruno, 46
Lefebvre, Henri, 47
Lewis, Saskia, 125, 130
lion dance, 130–32, 132*fig.*, 143

"Listen to the Voice of the Homeland"
    (Kim Cúc), 73
listening practices, 21; active, 54–56; anticipatory,
    55; buildings and, 130; collective, 22, 79;
    guest, 138, 173n2; hungry, 138–39; identity
    linked to, 81–82; individualized, 72; to *loa
    phương*, 79, 81, 84; medical, 20, 25–28;
    participant, 22–23; quarantining and,
    54; relational, 62, 68; sonic color lines
    reinforced by, 81; across sonic ecologies,
    128; sounding practices linked to, 78
Littlefield, David, 125, 130
live virus trackers, 33
loa phương (public loudspeaker), 139; artistic
    representations of, 75, 76*fig.*; as chroniclers
    of history, 75; criticisms of, 74, 81–82;
    deactivation, 162n24; debates over, 81–82;
    identity and, 81–82; impact of, in public
    sphere, 62–63; listening practices, 79, 81,
    84; live transmissions, 82; as notice boards,
    80; as panaudicon, 70–72; in Quang Trung
    housing complex, 72*fig.*; remediation of,
    82–83; time marked by, 70–71; top-down
    communication via, 79–80; vintage
    loudspeakers as commemorative objects,
    74–75; in wartime, 74
lockdown, COVID-19, 2*fig.*, 22, 60, 80–81,
    175n15; aftermath of, 111; defiance of, 90;
    Delta variant, 86; responsibility during,
    91–92
Lombardy, Italy, 64
loudspeakers. See *loa phương*

Maierhofer-Lischka, Margarethe, 74
Marr, David, 19, 162n28
masking, 12, 43, 156n14
masks: ATMs, 45, 143; donations of, 11; export
    ban on, 45
massage chain, 111*fig.*
Mattern, Shannon, 73, 121
Mattingly, Cheryl, 41
McCain, John, 1, 119
MDRI. *See* Mekong Development Research
    Institute
media convergences, 82
medical declarations, 37, 50, 80, 158n26.
    See also health e-declarations
medical listening techniques, 20, 25–28
medical sounding practices, 20
Mekong Development Research Institute
    (MDRI), 84, 144, 156n6
memes, 34

Metropole Hotel, 135
microphones, 139
Mid-Autumn Festival, 131, 132*fig.*, 140–41
migrant workers, 23, 61, 133–36
Minh Beta, 103
Ministry of Culture, Sports, and Tourism, 125
Ministry of Education and Training (MOET), 46
Ministry of Health, Vietnam, xiii, 29–30, 33,
    103, 126; Department of Preventative
    Medicine, 131
Ministry of Information and Communications,
    45, 77*fig.*
mismanagement, 10–12
Moderna (COVID-19 vaccine), 145
MOET. *See* Ministry of Education and Training
Mol, Annemarie, 16
Molave typhoon, 173n7
moral framing: of citizenship, 77; of outbreak
    narratives, 3–4
Morning Coffee, 5
multitasking, 81
mutual aid, 23, 45, 145–46, 157n20
#MuzikDapDich, 146
Myung-Ok Lee, Marie, 3

Nancy, Jean-Luc, 54
National Humanitarian Portal 1400, 45
National Steering Committee for COVID-19
    Prevention and Control, 64, 78, 84, 145,
    156n7, 168n9; infection rate updates from, 79
National Technical Regulation on Noise, 165n44
nationalism, 64, 65*fig.*, 81, 149n4; through song,
    xv, 19, 81, 149n4
NCOVI (app), 33
neoliberalism, 42, 124
the new normal, 93, 97, 103, 127; sounding,
    130–32
Ng, Stephanie, 134
Ngũ Xã Island, 66, 108*fig.*
người bảo vệ (security guard), 24–28
người giúp việc (housekeeper), 57–61
Nguyễn Thế Sơn, 75
Nguyễn Xuân Phúc, 64
Nội Bài Airport, 36; as COVID-19 test site, 37*fig.*
noise, 86, 87, 164n43; as dissent, 91–92
Noise (Attali), 84
nonadherence, 87, 140
noncompliance, 52, 86, 92
nonessential workers, 23, 94–97
nonhuman life, 116*fig.*; human-nonhuman
    animal encounters, 101*fig.*, 114–18; sonic
    ecologies of, 115*fig.*

normalcy, sounding, 90–93
nostalgia, 74–75
notice board (*bảng thông báo*), 80, 163n34

ocularcentrism, 16, 54, 107–8
OECD. *See* Organization for Economic Cooperation and Development
Old Quarter (Phố cổ), Hanoi, 60–61, 109, 129
Ông Địa (God of Earth), 132
#ONhaLaYeuNuoc, 64
OPAs. *See* overseas performing artists
Organization for Economic Cooperation and Development (OECD), 173n8
Orientalism: in Cold War, 10; COVID-19 pandemic and, 3; in outbreak narratives, 3
Otherness, 10, 11
outbreak narratives: gender and, 3, 160n3; moral framing of, 3–4; Orientalism in, 3
overseas performing artists (OPAs), 134

panaudicon, *loa phường* as, 70–72
pandemic precarity, 24, 90, 127
paperwork, 49–50
participant listening, 15, 22–23, 91
Patient 17, xx, 1–2, 3, 8, 104, 160n3; containment strategy, 4–5; public shaming of, 4, 160n3
Patient 91, 160n3
Patient 1342, 167n5
PCR tests. *See* polymerase chain reaction tests
People's Committee. *See* Ward People's Committee
performance: of compliance, 84–87; Goffman on, 86
personal protective equipment (PPE), xvii, 42, 158n28; availability of, 44
Peterson, Marina, 164n43
Philippine singer, 133–36
Plan 1749/KH-BVHTTDL, 125
plastic waste, 10, 39, 156n9, 166n52
polymerase chain reaction (PCR) tests, 51
portable speakers, 83*fig.*, 121
power dynamics, in COVID-19 pandemic, 32
PPE. *See* personal protective equipment
privacy, 34–35
property rights, 87
Proschan, Frank, 68
protest songs, 88
Przybylski, Liz, 13–14
public ear, 16
public loudspeaker. See *loa phường*

public sphere: impact of *loa phường* in, 62–63; sonic socialism and, 73; sounding practices in, 62–63

QR codes, 38
Quang Trung housing complex, 72*fig.*
Quarantine Pledge, 49
quarantining, 22, 27, 46–47, 99–100; care and control in, 48–56; cohorts, 5; listening and, 54
quiet soundings, 129–30

racial discrimination, xvii, 107, 134
rap music, Vietnamese, 141, 146
reciprocal care relations, 111*fig.*
The Red Days III, 76*fig.*
red songs, 81
relational attunements, 67–68, 124–25
relational listening, 62, 68
Residential Quarantine Guidelines, 49
Resolution 20/NQ-CP, 157n15
Resolution 42/NQ-CP, 166n3
Resolution 52-NQ/TW, 35, 155n5
responsibility, 30, 79, 105; collective, 18, 140*fig.*, 146; during lockdown, 91–92; sonic governance of care and, 32–33
rice ATMs, 45
ride-hailing apps, 94–95
Robinson, Dylan, 138–39, 173n2

Sa Pa District, 126
Sahlins, Marshall, xx
SARS. *See* severe acute respiratory syndrome
scarcity, 40, 41, 42
Scarry, Elaine, 32, 34, 110
Schafer, R. Murray, 151n1
securitization, 27, 35–37
security guard (*người bảo vệ*), 24–28, 43, 58; public safety rituals and, 48
selfies, 34
Seventeen Saloon, 134
severe acute respiratory syndrome (SARS), 152n5; in Vietnam, 9–10
shelter-in-place orders, 63, 67–68, 92, 166n52
shifting vocalizations, 121*fig.*
shortage paradigm, 41; in United States, 42
sidewalk checks, 118–20, 169n23
Siisiäinen, Lauri, 15
silence, 67–68, 112, 128
SIM cards, 29
Sinopharm (COVID-19 vaccine), 145
Six Senses Resort, 134–35
smartphones, 30–31, 33, 39

social distancing. *See* spatial distancing

socialism, 10; disaster socialism, 22, 23, 64, 145; everyday socialism, 41; market socialism, 143; socialist paternalism, 142; socialist sounding, 73

soft authority, 30

solidarity economy, 44–48

Solnit, Rebecca, 106

the sonic, 8–9; defining, 7

sonic agency, 17, 84, 169n13

sonic anarchiving, 112–14, 173n4

Sonic Bodies (Henriques), 150n5

sonic campaigns, 77–82; defining, 78

sonic color lines, 81

sonic convergences, 82–84, 83*fig.*

sonic cues, 15, 20, 58, 119; as indicators of disease, xiii, 79; orientation via, 69; temporality and, 69–70

sonic dissent, 87–88, 91*fig.*, 103; noise as, 91–92

sonic distortions, 138–43. *See also* distortion

sonic dominance, xvi, 17, 22, 141

sonic ecologies: interactivity of, 110–11; listening across, 128; mutation of, 113–14; nonhuman, 115*fig.*

sonic governance of care, 9, 15–18, 147; defining, 16; responsibility and, 32–33

sonic socialism, 18–21, 63, 73–77, 137–38, 162n27; defining, xvi; digitalization in, 82; public sphere and, 73

sonic sociality, 88–90

sonic trade, 120–23

sonosphere, 159n32

Sontag, Susan, 4, 149n3, 150n4

sound: ambient, 67–68; Blackness and, 15; of construction, 90–91; in COVID-19 pandemic, 17; cultural values transmitted by, 15; mapping, 112–14, 113*fig.*; vision versus, 8–9. *See also specific topics*

sounding practices, 8; collective, 22; defining, 150n5; listening practices and, 78; medical, 20; of the new normal, 130–32; of normalcy, 90–93; in public sphere, 62–63; quiet, 129–30; socialist, 73; state power and, 79, 84

soundmarks, 67, 70, 112

soundscape, defining, 151n1

soundwalking, 93, 100–101, 106–9; as Black feminist practice, 109; in ghost towns, 125–32; lakeshore contours, 110–25

soundwork, defining, 15

Space Speaker, 146

spatial distancing, 22–23, 28, 42, 79, 98, 125, 134; in highlands, 161n17; nonadherence to,

120, 131; temporality of, 110; "Việt Nam Ơi! Đánh Bay Covid" during, 106

speakers, portable, 82, 83*fig.*, 123*fig.*

spectral uncertainty, 130

Springgay, Stephanie, 100, 111, 112

state power: disaster governance and, 8; emergency conditions and, 32; sounding practices and, 79, 84

staycations, 126, 135

Sterne, Jonathan, 139, 155n2

stillness, 67–68

street art, 140*fig.*

street vendors, 43, 81, 121; disappearance of, 66, 102; iconic calls of, 23, 43, 60, 120; policing of, 119; return of, 120–22, 122*fig.*

supermarkets, 43

superspreaders, 1–2, 23, 150n2; stigmatization of, 3–4

surveillance, 17; medical, 5, 26. *See also* syndromic surveillance

sweeping, 59*fig.*, 111

Symbiotic Café, 75*fig.*

syndromic surveillance, 33

Tạ Hiện "Beer Street" neighborhood, 128–29, 128*fig.*, 130

Taipei, 29

T.B.O., 141

Tedros Adhanom Ghebreyesus, xviii, 150n6

temperature screening, 26, 43, 52–53

temporality: auditory cues and, 69–70; of COVID-19 pandemic, xx, 69–70, 131; *loa phường* marking time, 70–71; of social distancing, 110

testing sites, COVID-19, 37*fig.*

Thanh Niên Road, 66, 102, 119

Thinking in an Emergency (Scarry), 110

Thông điệp 5K (5k Message) Campaign, 131

thresholds, 111–12, 129–30

Thủy Trung Tiên Temple, 119

TikTok, xvi

total society quarantine, 58, 62–63

tourism, 125–26, 134

Trnka, Susanna, 46

Tronto, Jane, 16

Trúc Bạch Lake, 100, 106, 117, 122*fig.*, 129; walking path, 101*fig.*

Trúc Bạch Ward, 78, 88; health screening in, 52*fig.*; medical station, 50, 80; outbreak, xx, 1–2, 5, 7–8; urban caretaking in, 58

Truman, Sarah E., 100, 111, 112

Trundle, Catherine, 46

Trường Sa Island, 65*fig.*
Trường Sơn veterans, 45
Turner, Victor, 112
Typhoid Mary, 3

United Nations Development Programme
    (UNDP), 84, 156n6
United States: response to COVID-19 pandemic
    in, xviii, 11, 141–42; shortage paradigm in,
    42; Vietnam's relationship with, 42, 152n9;
    war in Vietnam, 74, 153n13
University of California at Riverside, xiii
urban caretaking, 39–40, 58, 59
urban ocularcentrism, 9
urban specters, 127–29

vaccines, COVID-19, 145, 157n19, 174n10
Verdery, Katherine, 143
"Việt Nam Ơi! Đánh Bay Covid" (Hey Vietnam!
    Let's Fight COVID), 81, 103–6, 168n11;
    patriotic sentiments of, 105–6; prevalence
    of, during social distancing, 106; video for,
    104–5, 105*fig.*
Vietnam: acoustic traditions in, xv–xvi, 18; avian
    flu in, 9–10; as country of planning, 141–43;
    decolonization of, 19, 73; digitalization in, 39;
    disaster governance in, 35–36; exceptionalism
    in, 85; public health infrastructure in, 10,
    22, 152n6; response to COVID-19 pandemic
    in, xviii, xx, 141; revolution and building of
    socialism, 125; SARS in, 9–10; United States
    relationship with, 42, 152n9
Vietnam National University, 145
Vietnam syndrome, 11

Vietnamese Television (VTV), 5
Vinastalgia, 74
Vinh City, 72*fig.*
virus, as invading enemy, 64, 77, 149n3
vocalizations, 121*fig.*
Voegelin, Salomé, 67, 86
Voice of Vietnam, 81
volunteerism, 45
VTV. *See* Vietnamese Television
Vũ Đức Đam, 78
Vũ Trọng Phung, 18, 154n29, 157n23

Wald, Priscilla, 3
walking, 168n8; as multisensory methodology,
    107; walking-notes, 66; walking-with,
    107–8. *See also* footwork; soundwalking
Walking While Black (Cadogan), 106
Wanderlust (Solnit), 106
Ward People's Committee, 5, 102, 123*fig.*, 162n30;
    public broadcasts, 70, 80
West Lake, 117, 140*fig.*, 169n20
Westerkamp, Hildegard, 117
wet marketplaces, 42–43
WHC. *See* World Health Organization
Women's Day, Vietnamese, 125
World Climate Summit, 173n7
World Health Organization (WHO), xvi, 5
Wowy, 146
Wuhan, xx, 3, 149n3

Yến Năng, 75

Zalo, 33, 90, 155n2
zoonotic epidemics, 35. *See also specific topics*

Founded in 1893,
UNIVERSITY OF CALIFORNIA PRESS
publishes bold, progressive books and journals
on topics in the arts, humanities, social sciences,
and natural sciences—with a focus on social
justice issues—that inspire thought and action
among readers worldwide.

The UC PRESS FOUNDATION
raises funds to uphold the press's vital role
as an independent, nonprofit publisher, and
receives philanthropic support from a wide
range of individuals and institutions—and from
committed readers like you. To learn more, visit
ucpress.edu/supportus.